GAMBLING IN AMERICA

ISSN 2373-3861

GAMBLING IN AMERICA

Stephen Meyer

INFORMATION PLUS® REFERENCE SERIES
Formerly Published by Information Plus, Wylie, Texas

Farmington Hills, Mich • San Francisco • New York • Waterville, Maine
Meriden, Conn • Mason, Ohio • Chicago

Gambling in America

Stephen Meyer
Kepos Media, Inc.: Steven Long and Janice Jorgensen, Series Editors

Project Editor: Laura Avery

Rights Acquisition and Management: Ashley M. Maynard, Carissa Poweleit

Composition: Evi Abou-El-Seoud, Mary Beth Trimper

Manufacturing: Rita Wimberley

Product Design: Kristin Julien

Gale
27500 Drake Rd.
Farmington Hills, MI 48331-3535

ISBN-13: 978-0-7876-5103-9 (set)
ISBN-13: 978-1-4103-2548-8

ISSN 2373-3861

This title is also available as an e-book.
ISBN-13: 978-1-4103-3271-4 (set)
Contact your Gale sales representative for ordering information.

Printed in the United States of America
1 2 3 4 5 22 21 20 19 18

TABLE OF CONTENTS

PREFACE vii

CHAPTER 1

Gambling in the United States: An Overview 1

This chapter traces the history of gambling from ancient and medieval times through 19th-century Europe and the United States to the modern era, including 21st-century developments. It concludes with a look at social and practical issues surrounding this topic and a report of public opinion on the moral acceptability of gambling.

CHAPTER 2

Supply and Demand: Who Offers Gambling? Who Gambles? 9

Corporations, small businesses, criminals, charities, and governments profit from offering gambling opportunities. Profiles of each appear in this chapter, which also addresses the participation rates, favorite wagering activities, and demographics of American gamblers. A discussion of gambling addiction closes the chapter.

CHAPTER 3

An Introduction to Casinos 21

Casinos may be as small as a single room or as large as a megaresort, but they all offer games of chance and skill played at tables or machines. The historical and current status of casinos, their social acceptability, the wide variety of games they offer, and the types of gamblers they attract are examined in this chapter.

CHAPTER 4

Commercial Casinos 25

Casinos owned and operated by large and small companies are the focus of this chapter, which outlines the history and current status of casinos in the states in which they are legal. Legislative measures authorizing and regulating these establishments, plus the economic impact casinos have on these states, are described in detail.

CHAPTER 5

Native American Tribal Casinos 45

Gaming establishments operated by Native American tribes have spread throughout the United States since Congress passed the Indian Gaming Regulatory Act in 1988. As explained in this chapter, these casinos can be legally operated only by officially recognized tribes, and they are subject to regulation by federal, state, and tribal governments.

CHAPTER 6

The Economic and Social Effects of Casinos 57

Although views of the positive and negative effects of casinos are inherently subjective and subject to dispute among experts, this chapter offers a general assessment. It considers how casinos affect economics, crime, suicide rates, personal bankruptcy, compulsive gambling, underage gambling, and politics.

CHAPTER 7

Lotteries 73

U.S. lotteries are operated by state governments to raise public funds. This chapter reports on the types of games offered (such as lotto, instant, scratch-off, and second-chance), multistate games (such as Powerball and Mega Millions), lottery players, the role of retailers in selling tickets, and the economic and social effects of lotteries.

CHAPTER 8

Sports Gambling 87

Three types of sports gambling exist in the United States: pari-mutuel betting on horse and greyhound races and on the ball game jai alai, betting through a bookmaker, and illegal wagering. Each is illustrated in detail and assessed in terms of positive and negative effects.

CHAPTER 9

Internet Gambling. 101

Internet gambling represents a new frontier for the gambling industry, for federal and state government regulators, and for U.S. residents. This chapter delves into the evolving legal and commercial fortunes of Internet gambling in the United States, including the pioneering legalization of the activity in Delaware, Nevada, and New Jersey.

IMPORTANT NAMES AND ADDRESSES. 111

RESOURCES. 117

INDEX 119

PREFACE

Gambling in America is part of the *Information Plus Reference Series*. The purpose of each volume of the series is to present the latest facts on a topic of pressing concern in modern American life. These topics include the most controversial and studied social issues of the 21st century: abortion, capital punishment, crime, the economy, education, health care, immigration, national security, race and ethnicity, social welfare, women, youth, and many more. Although this series is written especially for high school and undergraduate students, it is an excellent resource for anyone in need of factual information on current affairs.

By presenting the facts, it is the intention of Gale, a Cengage Company, to provide its readers with everything they need to reach an informed opinion on current issues. To that end, there is a particular emphasis in this series on the presentation of scientific studies, surveys, and statistics. These data are generally presented in the form of tables, charts, and other graphics placed within the text of each book. Every graphic is directly referred to and carefully explained in the text. The source of each graphic is presented within the graphic itself. The data used in these graphics are drawn from the most reputable and reliable sources, such as from the various branches of the U.S. government and from private organizations and associations. Every effort has been made to secure the most recent information available. Readers should bear in mind that many major studies take years to conduct and that additional years often pass before the data from these studies are made available to the public. Therefore, in many cases the most recent information available in 2017 is dated from 2014 or 2015. Older statistics are sometimes presented as well, if they are landmark studies or of particular interest and no more-recent information exists.

Although statistics are a major focus of the *Information Plus Reference Series*, they are by no means its only content. Each book also presents the widely held positions and important ideas that shape how the book's subject is discussed in the United States. These positions are explained in detail and, where possible, in the words of their proponents. Some of the other material to be found in these books includes historical background, descriptions of major events related to the subject, relevant laws and court cases, and examples of how these issues play out in American life. Some books also feature primary documents or have pro and con debate sections that provide the words and opinions of prominent Americans on both sides of a controversial topic. All material is presented in an evenhanded and unbiased manner; readers will never be encouraged to accept one view of an issue over another.

HOW TO USE THIS BOOK

Gambling has long been a favorite pastime worldwide, and its history in the United States dates back to the founding of the nation. It has been estimated that Americans spend more than $1 trillion per year on charitable gambling, betting on horse and greyhound races, lottery purchases, casino wagering, and other legal and illegal gambling activities. Much controversy surrounds the gambling industry. While pro-gambling elements argue that the economic benefits of gambling far outweigh any potential risks, some individuals cast doubt on its economic benefits, oppose gambling on moral grounds, or argue that it can cause an increase in various types of social problems. This book presents in-depth information on how casino gambling, sports gambling, lotteries, and Internet gambling work, provides up-to-date financial data for each, and addresses the effects of these and other gambling activities on the communities in which they take place. Also profiled are American attitudes toward gambling.

Gambling in America consists of nine chapters and three appendixes. Each chapter is devoted to a particular aspect of gambling in the United States. For a summary of the information that is covered in each chapter, please see the synopses that are provided in the Table of Contents. Chapters generally begin with an overview of the basic facts and background information on the chapter's topic, then proceed to examine subtopics of particular interest. For example, Chapter 4: Commercial Casinos offers a broad survey of casino gambling in the United States. A history of gambling in Nevada follows, with a focus on the emergence of the casino industry in Las Vegas. The chapter proceeds with overviews of casino gaming in individual states, providing detailed information on local gambling laws, annual taxation and revenues, amounts wagered, employment and attendance figures, types of games offered at various casinos, and other pertinent data. Readers can find their way through a chapter by looking for the section and subsection headings, which are clearly set off from the text. They can also refer to the book's extensive Index, if they already know what they are looking for.

Statistical Information

The tables and figures featured throughout *Gambling in America* will be of particular use to readers in learning about this topic. These tables and figures represent an extensive collection of the most recent and valuable statistics on gambling, as well as related issues—for example, the sales and profits of state lotteries, the demographics of certain gamblers, 20 questions designed to determine whether a person is a compulsive gambler, and the growth of tribal gaming revenues. Gale, a Cengage Company, believes that making this information available to readers is the most important way to fulfill the goal of this book: to help readers understand the issues and controversies surrounding gambling in the United States and reach their own conclusions.

Each table or figure has a unique identifier appearing above it, for ease of identification and reference. Titles for the tables and figures explain their purpose. At the end of each table or figure, the original source of the data is provided.

To help readers understand these often complicated statistics, all tables and figures are explained in the text. References in the text direct readers to the relevant statistics. Furthermore, the contents of all tables and figures are fully indexed. Please see the opening section of the Index at the back of this volume for a description of how to find tables and figures within it.

Appendixes

Besides the main body text and images, *Gambling in America* has three appendixes. The first is the Important Names and Addresses directory. Here, readers will find contact information for a number of government and private organizations that can provide further information on different aspects of gambling. The second appendix is the Resources section, which can also assist readers in conducting their own research. In this section, the author and editors of *Gambling in America* describe some of the sources that were most useful during the compilation of this book. The final appendix is the Index. It has been greatly expanded from previous editions and should make it even easier to find specific topics in this book.

COMMENTS AND SUGGESTIONS

The editors of the *Information Plus Reference Series* welcome your feedback on *Gambling in America*. Please direct all correspondence to:

Editors
Information Plus Reference Series
27500 Drake Rd.
Farmington Hills, MI 48331-3535

CHAPTER 1

GAMBLING IN THE UNITED STATES: AN OVERVIEW

Gambling is an activity in which something of value is risked on the chance that something of greater value might be obtained, based on the uncertain outcome of a particular event. Organized gambling has become an industry because so many people are willing and even eager to risk their money in exchange for a chance at something bigger and better. The elements of risk and uncertainty actually add to gambling's appeal—and to its danger. Throughout history, various cultures have considered gambling alternately harmless and sinful, respectable and corrupt, and legal and illegal. Societal attitudes are dependent on customs, traditions, religion, morals, and the context in which gambling occurs.

Lawmakers have struggled to define gambling and determine which activities should and should not be legal. For example, betting activities with an element of skill involved (such as picking a horse in a race or playing a card game) might be more acceptable than those based entirely on chance (such as spinning a roulette wheel or playing slot machines). Acceptability also depends on who profits from gambling. Bingo games held for charity and lotteries that fund state programs have historically faced fewer legal restrictions than casinos run for corporate profit.

People who gamble are almost certain, in statistical terms, to lose money over the long term. The American Gaming Association (AGA), an advocacy group for the casino industry, notes in "The House Advantage: A Guide to Understanding the Odds" (2011, http://coloradogaming.com/wp-content/uploads/2015/11/odds_brochure_2011_final.pdf) that "it's important to remember that the house continues to have a statistical advantage in every play of every game, even against a skillful player." Casino games, such as lotteries and other forms of gambling, are designed so that the house (the casino or other sponsor of the game) is guaranteed to take in more money from players than it pays out. In certain games, personal skill can reduce the size of the house's advantage, but it cannot eliminate the house advantage.

So why do people gamble? Common sense suggests that risking something of value on an event in which the odds of success are intentionally stacked against the gambler is irrational. Psychologists and social scientists postulate a variety of reasons for gambling, including the lure of money, the excitement and fun of the activity, and the influence of peers. At its deepest level, gambling may represent a human desire to control the randomness that seems to permeate life. Whatever the drive may be, it must be strong, because an entire gambling culture has developed in the United States in which entrepreneurs (legal and otherwise) offer people opportunities to gamble, and business is booming.

THE HISTORY OF GAMBLING

Ancient Times

Archaeologists have discovered evidence that people in Egypt, China, Japan, and Greece played games of chance with dice and other devices as far back as 2000 BC. According to *Encyclopaedia Britannica*, loaded dice—which are weighted to make a particular number come up more often than others—have been found in ancient tombs in Egypt, the Far East, and even North and South America.

Dice are probably the oldest gambling implements known. They were often carved from sheep bones and known as knucklebones. They are mentioned in several historical documents, including the Mahabharata, the epic poem and philosophy text written in India approximately 2,500 years ago. A story in the New Testament of the Bible describes Roman soldiers throwing dice to determine who would get the robe of Jesus (c. 4 BC–AD 29). Roman bone dice have been found dating from the first to the third centuries AD. The Romans also gambled on chariot races, animal fights, and gladiator contests.

The Medieval Period

During medieval times (c. 500–c. 1500) gambling was legalized by some governments, particularly in areas of modern-day Germany, Italy, the Netherlands, and Spain. England and France were much less permissive, at times outlawing all forms of gambling. For example, King Louis IX (1215–1270) of France prohibited gambling during his reign for religious reasons. Still, illegal gambling continued to thrive.

During this period Christian powers in Europe launched the Crusades (military expeditions against Muslim powers that controlled lands considered holy by Christians). They also permitted gambling, but only by knights and people of higher rank. Violators were subject to severe whippings. Nevertheless, even among the titled gamblers there was a legal limit on how much money could be lost, a concept that later would come to be known as limited-stakes gambling.

English knights returned from the Crusades with long-legged Arabian stallions, which they bred with sturdy English mares to produce Thoroughbred racehorses. Betting on private horse races became a popular pastime among the nobility. Card games also became popular in Europe around the end of the 14th century. According to the International Playing-Card Society, in "History of Playing-Cards" (January 24, 2017, http://i-p-c-s.org/history.html), one of the earliest known references to playing cards in Europe dates from 1377. During the late 1400s and early 1500s lotteries were used in Europe to raise money for public projects. Danny Lewis notes in "Queen Elizabeth I Held England's First Official Lottery 450 Years Ago" (Smithsonianmag.com, January 13, 2016) that Queen Elizabeth I (1533–1603) established the first English state lottery in 1567.

Precolonial America and the Colonial Era

Native Americans played games of chance as part of tribal ceremonies and celebrations hundreds of years before the Europeans arrived in North America. One of the most common was a dice and bowl game in which five plum stones or bones carved with different markings were tossed into a bowl. Wagers were placed before the game began, and scoring was based on the combination of markings that appeared after a throw. The Cheyenne called the game *monshimout*. A similar game was called *hubbub* by the Arapaho and New England tribes.

European colonists brought gambling traditions with them to the New World. Historical accounts report that people in parts of New England gambled on horse racing, cockfighting, and bullbaiting. Bullbaiting was a blood sport in which a bull was tethered to a stake and attacked by dogs. The dogs were trained to torment the bull, which responded by goring the dogs. Spectators gambled on how many of the dogs the bull would kill.

In 1612 King James I (1566–1625) of England created a lottery to provide funds for Jamestown, Virginia, the first permanent British settlement in North America. Lotteries were later held throughout the colonies to finance the building of towns, roads, hospitals, and schools and to provide other public services.

Many colonists, however, disapproved of gambling. The Pilgrims and Puritans fled to North America during the 1620s and 1630s to escape persecution in Europe for their religious beliefs. They believed in a strong work ethic that considered labor morally redeeming and viewed gambling as sinful because it wasted time that could be spent on productive endeavors.

Cockfighting, bear- and bullbaiting, wrestling matches, and footraces were popular gambling sports throughout Europe during the 16th and 17th centuries. The predecessors of many modern casino games were also developed and popularized during this period. For example, the invention of the roulette wheel is often attributed to the French mathematician Blaise Pascal (1623–1662).

Gambling among British aristocrats became so customary during the early years of the 18th century that it presented a financial problem for the country. Gentlemen gambled away their belongings, their country estates, and even their titles. Cuthbert William Johnson noted in *The Law of Bills of Exchange, Promissory Notes, Checks, &c* (1839) that large transfers of land and titles were disruptive to the nation's economy and stability, so the reigning monarch, Queen Anne (1665–1714), responded in 1710 with the Statute of Anne, which made large gambling debts "utterly void, frustrate, and of none effect, to all intents and purposes whatsoever." In other words, large gambling debts could not be legally enforced. This prohibition prevailed in common law for centuries and is still cited in U.S. court cases. Queen Anne is also known for her love of horse racing, which became a popular betting sport (along with boxing) during her reign.

A surge of evangelical Christianity swept through England, Scotland, Germany, and the North American colonies during the mid- to late 1700s. Historians refer to this period as the Great Awakening, a time when conservative moral values became more prevalent and widespread. Evangelical Christians considered gambling to be a sin and dangerous to society, and religion became a powerful tool for bringing about social change.

In October 1774 the Continental Congress of the North American colonies issued the Articles of Association (2008, http://avalon.law.yale.edu/18th_century/contcong_10-20-74.asp), which stated in part that the colonists "will discountenance and discourage every species of extravagance and dissipation, especially all horse-racing, and all kinds of games, cock fighting, exhibitions of shews [*sic*], plays, and other expensive diversions and entertainments." The purpose of the directive was to "encourage frugality, economy, and industry."

The 19th Century

In general, gambling was tolerated as long as it did not upset the social order. According to James R. Westphal et al., in "Gambling in the South: Implications for Physicians" (*Southern Medical Journal*, vol. 93, no. 9, September 2000), Georgia, South Carolina, and Virginia passed versions of the Statute of Anne during the colonial period to prevent gambling from getting out of hand. New Orleans became a gambling mecca in the 1700s and 1800s, even though gambling was outlawed during much of that time. During the 1830s almost all southern states outlawed gambling in public places; some exceptions, however, were made for "respectable gentlemen."

In 1823, 11 years after becoming a state, Louisiana passed a licensing act that legalized several forms of gambling and licensed several gambling halls in New Orleans. Although this act was repealed in 1835, casino-type gambling continued to prosper and spread to riverboats traveling the Mississippi River. Professional riverboat gamblers soon developed an unsavory reputation as cheats and scoundrels. Several historians trace the popularization of poker and craps in the United States to Louisiana gamblers of that period. Riverboat gambling continued to thrive until the outbreak of the Civil War (1861–1865).

Andrew Jackson (1767–1845) was president of the United States from 1829 to 1837. The Jacksonian era, during which a wave of evangelical Christianity swept the country, was associated with renewed attention to social problems and a focus on morality. According to I. Nelson Rose of the Whittier Law School, in "The Rise and Fall of the Third Wave: Gambling Will Be Outlawed in Forty Years" (William R. Eadington and Judy A. Cornelius, eds., *Gambling and Public Policy: International Perspectives*, 1991), gambling scandals and the spread of a conservative view of morality led to an end to most legal gambling in the United States by the mid-1800s.

Across the country private and public lotteries were plagued by fraud and scandal and fell into disfavor. During the 1840s most southern states banned lotteries on moral grounds. By 1862 only two states, Missouri and Kentucky, had legal lotteries. Lotteries, however, were reinstated after the Civil War to raise badly needed funds. In 1868 Louisiana implemented a lottery known as the Serpent. Although it was extremely popular, the lottery was plagued with fraud and was eventually outlawed by the state in 1895. Casino gambling, which had been legalized again in Louisiana in 1869, was outlawed at the same time as the lottery.

Frontier gambling in the Old West, both legal and illegal, peaked during the mid- to late 19th century. Saloons and other gambling houses were common in towns that catered to cowboys, traders, and miners. Infamous gamblers of the time included Doc Holliday (1851–1887), Bat Masterson (1853–1921), Alice Ivers Tubbs (1851–1930), and Wild Bill Hickok (1837–1876). Hickok was shot while playing poker in 1876. At the time, he held a hand of two black aces and two black eights, which came to be known as the "dead man's hand."

Gambling fell into disfavor as the 19th century ended. In England, Queen Victoria (1819–1901) ruled from 1837 to 1901; her rule was characterized by concern for morality and by the spread of conservative values. These attitudes permeated American society as well. Gambling fell out of favor as a pastime for respectable people. Many eastern racetracks and western casinos were pressured to close for moral and ethical reasons. As new states entered the Union, many included provisions against gambling in their constitutions. By federal law, all state lotteries were shut down by 1900.

GAMBLING IN THE UNITED STATES SINCE 1900

At the dawn of the 20th century, there were 45 states in the Union. The territories of Arizona, New Mexico, and Oklahoma gained statehood between 1907 and 1912. According to Rose, the closure of casinos in Arizona and New Mexico was a precondition for statehood. In 1910 Nevada outlawed casino gambling. That same year horse racing was outlawed in New York, and almost all gambling was prohibited in the United States. The only legal gambling options at the time were horse races in Maryland and Kentucky and a few isolated card clubs.

Legalized Casinos in Nevada

The 1930s were a time of reawakening for legal gambling interests. Many states legalized horse racing and charitable gambling. Nevada went even further. In 1931 its legislature made casino gambling legal again. It seemed like a logical step: frontier gambling was widely tolerated in the state, although gambling was officially illegal. More important, Nevada, like the rest of the country, was suffering the effects of the Great Depression (1929–1939), and it sought to cash in on two events. The state's divorce laws were changed during the early 1930s to allow the granting of a divorce after only six weeks of residency, so people from other states temporarily moved into small motels and inns to satisfy the residency requirement. At the same time, construction began on the massive Boulder Dam (which would later be renamed Hoover Dam), only 30 miles (48 km) to the east of Las Vegas. Thousands of construction workers—much like the people waiting for their divorces to become final—were all potential gamblers.

Small legal gambling halls opened in Reno (in the northern part of the state), but they catered mostly to cowboys and local residents and had a reputation for being raunchy and wild. In April 1931 the first gambling licenses were issued in Las Vegas. The first big casino, El Rancho Vegas, was opened in 1941 on what would later be known as the Strip.

Many in the business world doubted that casino gambling in Nevada would be successful. Most of the casino hotels were small establishments operated by local families or small private companies, and some were dude ranches (western-style resorts that offered horseback riding). They were located in hot and dusty desert towns far from major cities, had no air conditioning, and offered few amenities to travelers. Also, there was no state oversight of gambling activities.

However, the end of Prohibition—which had made it illegal to import or sell alcoholic beverages in the United States—brought another element to Las Vegas. During the Prohibition Era (1920–1933) organized crime syndicates operated massive bootlegging rings and became very powerful and wealthy. When Prohibition ended, they switched their focus to gambling. Organized criminals in New York and Chicago were among the first to see the potential of Nevada. Meyer Lansky (1902?–1983) and Frank Costello (1891–1973) sent fellow gangster Bugsy Siegel (1906–1947) west to develop new criminal enterprises. Siegel invested millions of dollars of the mob's money in a big and lavish casino in Las Vegas that he was convinced would attract top-name entertainers and big-spending gamblers. The Flamingo Hotel and Casino opened in 1946. It was a failure at first, and Siegel was soon killed by his fellow mobsters.

POST–WORLD WAR II. Nevada's casinos grew slowly until after World War II (1939–1945). Postwar Americans were full of optimism and had spending money. Tourism began to grow in Nevada. Las Vegas casino resorts attracted Hollywood celebrities and famous entertainers. The state began collecting gaming taxes during the 1940s. The growing casinos in Las Vegas provided good-paying jobs to workers who brought their families with them, building a middle-class presence. In 1955 the state legislature created the Nevada Gaming Control Board within the Nevada Tax Commission. The Nevada Gaming Commission was established four years later.

CORPORATE GROWTH. The Las Vegas casinos continued to grow during the 1960s. By that time, organized crime syndicates used respectable front men in top management positions while they manipulated the businesses from behind the scenes. Publicly held corporations had been largely kept out of the casino business by a provision in Nevada law that required every individual stockholder to be licensed to operate a casino.

One corporation that was able to get into the casino business was the Summa Corporation, a spin-off of the Hughes Tool Company, which had only one stockholder: Howard Hughes (1905–1976). Hughes was a wealthy and eccentric businessman who owned the very profitable Hughes Aircraft Company. He spent substantial time in Las Vegas during the 1940s and 1950s and later moved there. In 1966 he bought the Desert Inn, a casino hotel on the Strip in Las Vegas. Later, he bought the nearby Castaways, Frontier, Sands, and Silver Slipper casinos.

Legend has it that mobsters threatened Hughes to drive him out of the casino business in Las Vegas, but he refused to leave. He invested hundreds of millions of dollars in Las Vegas properties and predicted that the city would be an entertainment center by the end of the century.

In 1967 the Nevada legislature changed the law to make it easier for corporations to own casinos. To combat organized crime, federal statutes against racketeering (the act of extorting money or favors from businesses through the use of intimidating tactics or by other illegal means) were enacted in 1971, and Nevada officials overhauled the casino regulatory system, making it more difficult for organized crime figures to be involved. Corporations and legitimate financiers began investing heavily in casino hotels in Las Vegas and in other parts of the state. In "Mob Ties" (LasVegasSun.com, May 15, 2008), Ed Koch and Mary Manning report that by the mid-1980s the link between organized crime and casino gambling in Las Vegas had significantly declined.

The Development of Gambling beyond Nevada

During the early 1970s the U.S. Commission on the Review of the National Policy toward Gambling studied Americans' attitudes about gambling and their gambling behavior. The commission found that 80% of Americans approved of gambling and that 67% engaged in gambling activities. In its final report, *Gambling in America* (1976), the commission concluded that state governments that were considering the legalization of gambling should set gambling policy without interference from the federal government, unless problems developed from the infiltration of organized crime or from conflicts between states.

In 1978 the first legal casino hotel outside of Nevada opened in Atlantic City, New Jersey. By the mid-1990s nine additional states had legalized casino gambling: Iowa (1989), South Dakota (1989), Colorado (1990), Illinois (1990), Mississippi (1990), Louisiana (1991), Indiana (1993), Missouri (1993), and Michigan (1996). As states saw the opportunity to generate revenue that could be used to fund various public priorities, the trend toward legalization accelerated. The AGA indicates in *2016 State of the States: The AGA Survey of the Casino Industry* (November 2016, https://www.americangaming.org/sites/default/files/2016%20State%20of%20the%20States_FINAL.pdf) that by 2015, 24 states allowed for the operation of commercial casinos.

Meanwhile, Native American tribes, whose lands are considered sovereign (independent) nations, moved into the casino industry with similar speed. Many tribes had long operated bingo halls to raise funds, and these became popular during the 1970s. Some of the most successful tribal bingo halls were high-stakes operations in Florida and Maine, where most other forms of gambling were prohibited. As the stakes grew, so did public scrutiny of the bingo halls, and the tribes began facing legal opposition from state governments. The tribes argued that their status as sovereign nations made them exempt from state laws against gambling. Tribes in various states sued, and the issue was debated in court for years.

Finally, the U.S. Supreme Court's landmark ruling in *California v. Cabazon Band of Mission Indians* (480 U.S. 202 [1987]) opened the door to tribal gaming when it found that gambling activities conducted on tribal lands did not fall within the legal jurisdiction of the state. In 1988 Congress passed the Indian Gaming Regulatory Act, which allowed federally recognized Native American tribes to open gambling establishments on their reservations if the states in which they were located already permitted legalized gambling.

In 2000 California voters passed Proposition 1A, which amended the state constitution to permit Native American tribes to operate lottery games, slot machines, and banking and percentage card games on tribal lands. Previously, in states that did not otherwise allow gambling, the tribes had been largely restricted to operating bingo halls. According to the National Indian Gaming Commission, in "Facts at a Glance" (April 6, 2017, https://www.nigc.gov/images/uploads/NIGC%20Uploads/aboutus/2016FactSheet-web.pdf), by 2015 tribal gaming was allowed and present in 28 states.

By the second decade of the 21st century, there were few populous areas of the United States where residents did not have ready access to a commercial or tribal casino. Besides commercial and tribal casinos, many states allow other forms of gambling in casinos or casino-like locations. Some states that do not have casinos with a full range of gaming machines and table games do allow for betting on horse or dog racing, and many such states allow racetracks to offer additional gambling options such as slot machines. These casino-like establishments are called racinos. Other states, such as Montana and California, allow for card or poker rooms, where players compete against one another rather than the house. Some states also offer video poker and other electronic gaming devices in retail locations other than casinos.

Despite the steady growth of the commercial casino industry, illegal gambling remained pervasive in the second decade of the 21st century. The AGA reports in the press release "New Report Shows Strong Ties between Illegal Gambling and Organized Crime" (September 30, 2015, https://www.americangaming.org/newsroom/press-releasess/new-report-shows-strong-ties-between-illegal-gambling-and-organized-crime) that in 2014 federal prosecutors convicted 80 illegal gambling operators in 23 states. According to the AGA, many of these operations had direct ties to organized crime.

STATE-SPONSORED LOTTERIES. In 1964 New Hampshire was the first state to legalize a modern state lottery. Called the New Hampshire Sweepstakes, it was tied to horse-race results to avoid laws that prohibited lotteries. New York established a lottery in 1967. Twelve other states followed suit during the 1970s. These legal lottery states were concentrated in the Northeast: New Jersey (1970), Pennsylvania (1971), Connecticut (1972), Massachusetts (1972), Michigan (1972), Maryland (1973), Illinois (1974), Maine (1974), Ohio (1974), Rhode Island (1974), Delaware (1975), and Vermont (1977).

An additional 23 states and the District of Columbia legalized lotteries during the 1980s and 1990s. The first multistate lottery game began operating in 1988 and included Iowa, Kansas, Oregon, Rhode Island, West Virginia, and the District of Columbia. It went through several incarnations before becoming the Powerball game in 1992. By 2017 seven more states had legalized lotteries, bringing the total to 44 states and the District of Columbia, and there was a second multistate game called Mega Millions that, along with Powerball, offered jackpots exceeding $500 million.

INTERNET GAMBLING. Internet gambling sites, many of them based in the Caribbean, began operating during the mid-1990s. Although some countries, such as the United Kingdom, embraced Internet gambling and began regulating the industry, the United States took action to interrupt online gambling activity by U.S. gamblers. Passage of the Unlawful Internet Gambling Enforcement Act of 2006 (UIGEA) made it illegal for banks and credit card companies to process payments from U.S. customers to gambling websites. Many websites immediately stopped accepting customers in the United States.

In September 2011, however, the U.S. Department of Justice issued an opinion that was viewed by many as a reversal of the UIGEA's prohibition on most types of online gambling. In *Whether Proposals by Illinois and New York to Use the Internet and Out-of-State Transaction Processors to Sell Lottery Tickets to In-State Adults Violate the Wire Act* (September 20, 2011, https://www.justice.gov/sites/default/files/olc/opinions/2011/09/31/state-lotteries-opinion_0.pdf), the Department of Justice indicates that federal law prohibits only online sports betting and that it is up to the states to determine the legality of other forms of online gambling. Delaware, Nevada, and New Jersey responded by legalizing online gambling and allowing for the operation of state-regulated gambling websites, which opened for business in 2013. In Nevada and New Jersey legal online gambling websites are operated by in-state companies, whereas in Delaware the state lottery runs the websites; only individuals physically located in these states' borders can register to play. As of 2017, this sector of the legal gambling industry was still in its infancy, but analysts expected that other states will eventually legalize Internet gambling. Meanwhile, illegally operating offshore sites continued to draw large numbers of online gamblers in the United States.

GAMBLING ISSUES AND SOCIAL IMPACT

In 1957 two men addicted to gambling decided to meet regularly to discuss the problems gambling had caused them and the changes they needed to make in their life to overcome it. After meeting for several months, each realized that the moral support offered by the other was allowing them to control their desire to gamble. They started an organization that was based on the spiritual principles used by Alcoholics Anonymous and similar groups to control addictions. The first group meeting of Gamblers Anonymous was held in September 1957 in Los Angeles, California.

As gambling became more widespread throughout the country, efforts were undertaken to help those whose lives had been negatively affected by gambling. In recognition of the wide social impact of the gambling industry, the American Psychiatric Association (APA) officially recognized pathological gambling as a mental health disorder in the third edition of the *Diagnostic and Statistical Manual of Mental Disorders* (*DSM-III*; 1980). Pathological gambling was listed under disorders of impulse control and described as a "chronic and progressive failure to resist impulses to gamble." During the 1980s many states began setting up programs to offer assistance to compulsive gamblers. Harrah's Entertainment became the first commercial casino company to officially address problem gambling when it instituted the educational campaigns Operation Bet Smart and Project 21 to promote responsible gaming and raise awareness about problems associated with underage gambling, respectively.

In 1996 Congress authorized the National Gambling Impact Study Commission to investigate the social and economic consequences of gambling in the United States. The federally funded group included nine commissioners representing pro- and antigambling positions. Existing literature was reviewed and new studies were ordered. The commission held hearings around the country at which a variety of people involved in and affected by the gambling industry testified. In June 1999 the commission concluded in *Final Report* (http://govinfo.library.unt.edu/ngisc/reports/fullrpt.html) that with the exception of Internet gambling, gambling policy decisions were best left up to state, tribal, and local governments. The commission also recommended that legalized gambling not be expanded further until all related costs and benefits were identified and reviewed.

That same year the National Academies Press published *Pathological Gambling: A Critical Review*, which identified and analyzed all available scientific research studies dealing with pathological and problem gambling. The researchers estimated that about 1.5% of American adults had been pathological gamblers at some point during their life, with about 1.8 million compulsive gamblers actively gambling during a given year. Although the researchers were able to draw some general conclusions about the prevalence of pathological gambling in the United States, they cited a lack of scientific evidence as a limiting factor in their ability to draw more specific conclusions. For example, they found that men were more likely than women to be pathological gamblers, but they lacked the data to estimate the prevalence of problem gambling among demographic subgroups such as the elderly or those with low incomes.

In the 2013 revision of the *DSM* (*DSM-5*), the APA reclassified pathological gambling as a gambling disorder, in recognition of the fact that the previous term was perceived by many to have pejorative (negative) connotations. The APA also moved the gambling diagnosis from the category Impulse-Control Disorders Not Elsewhere Classified to the category Substance-Related and Addictive Disorders, based on research demonstrating that it shares many common traits with disorders related to substance use. This reclassification is believed to offer possibilities for better recognition and treatment of the condition.

PUBLIC OPINION

Alan Mallach of the Brookings Institution suggests in *Economic and Social Impact of Introducing Casino Gambling: A Review and Assessment of the Literature* (March 2010, https://www.philadelphiafed.org/community-development/publications/discussion-papers/discussion-paper_casino-gambling.pdf) that "despite the increasing spread of gaming establishments across the United States and the large numbers of people that frequent them, Americans continue to be ambivalent about casinos." This ambivalence, Mallach notes, is evident in the fact that casinos operate within highly unusual legal and regulatory environments. For example, casinos in most states are taxed at far higher rates than other businesses, with the total tax burden for the gaming industry rising above 50% in some states. Additionally, many states require casinos to take the form of "riverboats" that are not actual boats but floating buildings. Such requirements are widely regarded as legal fictions reflecting an "out of sight, out of mind" mentality, whereby people who disapprove of gambling need not be squarely confronted with evidence

FIGURE 1.1

Public opinion on the moral acceptability of selected issues, 2017

[% morally acceptable (high or low points in trends are designated below)]

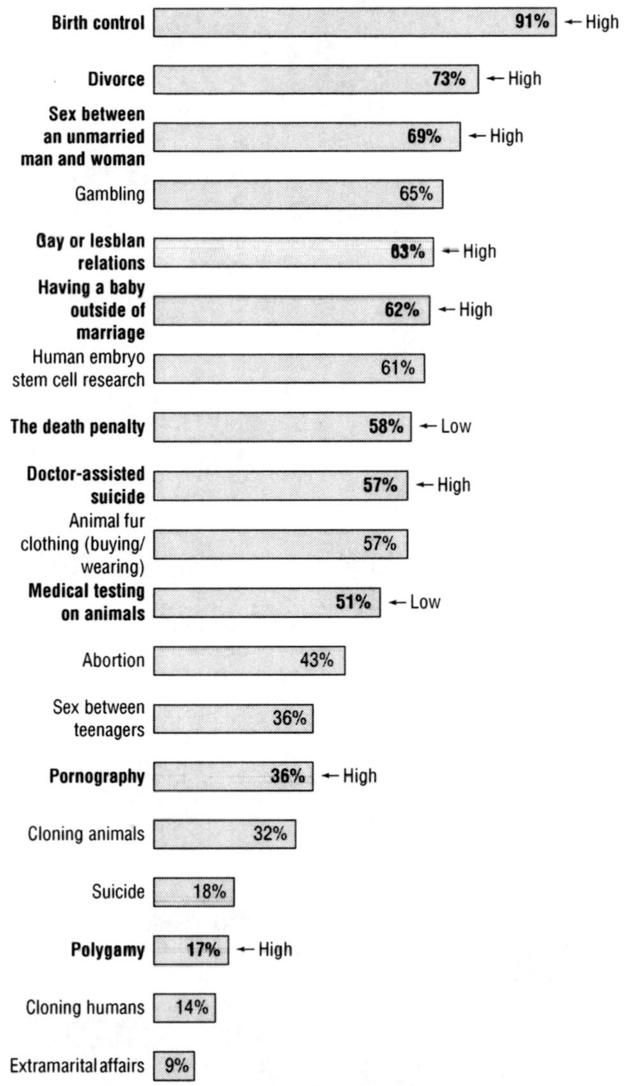

Issue	%	
Birth control	91%	← High
Divorce	73%	← High
Sex between an unmarried man and woman	69%	← High
Gambling	65%	
Gay or lesbian relations	63%	← High
Having a baby outside of marriage	62%	← High
Human embryo stem cell research	61%	
The death penalty	58%	← Low
Doctor-assisted suicide	57%	← High
Animal fur clothing (buying/ wearing)	57%	
Medical testing on animals	51%	← Low
Abortion	43%	
Sex between teenagers	36%	
Pornography	36%	← High
Cloning animals	32%	
Suicide	18%	
Polygamy	17%	← High
Cloning humans	14%	
Extramarital affairs	9%	

SOURCE: Jeffrey M. Jones, "Americans' Opinions about Moral Acceptability of Practices," in *Americans Hold Record Liberal Views on Most Issues*, The Gallup Organization, May 11, 2017, http://www.gallup.com/poll/210542/americans-hold-record-liberal-views-moral-issues.aspx?g_source=gambling&g_medium=search&g_campaign=tiles (accessed June 26, 2017). Copyright © 2017. Republished with permission of Gallup, Inc.; permission conveyed through Copyright Clearance Center, Inc.

TABLE 1.1

Percentage of Americans who gambled in the previous 12 months, by type of gambling, 2016

	Yes, have done this in past 12 months %
Bought state lottery ticket	49
Visited casino	26
Participated in office pool on the World Series, Super Bowl or other game	15
Bet on professional sports event	10
Other type of gambling	9
Played video poker machine	9
Bet on horse race	6
Played bingo for money	6
Bet on college sports event	5
Bet on boxing match	3
Gambled for money on the internet	3
Overall, have gambled in past 12 months	
Yes	64%

SOURCE: Zac Auter, "Gambling Behavior among U.S. Adults," in *About Half of Americans Play State Lotteries*, The Gallup Organization, July 22, 2016, http://www.gallup.com/poll/193874/half-americans-play-state-lotteries.aspx?g_source=gambling&g_medium=search&g_campaign=tiles (accessed June 26, 2017). Copyright © 2017. Republished with permission of Gallup, Inc.; permission conveyed through Copyright Clearance Center, Inc.

of its legality in their city or state. The uniquely restrictive approach to regulating casinos testifies to the fact that, as Mallach states, "many, if not most, communities would not entertain the idea of legalizing casino gambling were it not for the anticipated economic benefits."

In *Americans Hold Record Liberal Views on Most Moral Issues* (May 11, 2017, http://www.gallup.com/poll/210542/americans-hold-record-liberal-views-moral-issues.aspx), Jeffrey M. Jones of Gallup, Inc., reports that in 2017 nearly two-thirds (65%) of respondents expressed the opinion that gambling is morally acceptable. (See Figure 1.1.) Similar approval levels were found for gay or lesbian relations (63%), having a baby outside of marriage (62%), medical research using stem cells from human embryos (61%), and buying and wearing clothing made of animal fur (57%). Far fewer respondents expressed moral approval of married men and women having an affair (9%), cloning humans (14%), polygamy (17%), and suicide (18%), whereas a significantly larger proportion of those surveyed considered birth control morally acceptable (91%).

Zac Auter of Gallup reports in *About Half of Americans Play State Lotteries* (July 22, 2016, http://www.gallup.com/poll/193874/half-americans-play-state-lotteries.aspx) that in 2016, 64% of American adults had gambled at some point during the previous 12 months. (See Table 1.1.) Nearly half (49%) had purchased a state lottery ticket, while roughly a quarter (26%) had visited a casino.

CHAPTER 2

SUPPLY AND DEMAND: WHO OFFERS GAMBLING? WHO GAMBLES?

As with all businesses in a capitalist society, the gambling industry is driven by the principles of supply and demand. Gambling proponents argue that demand drives supply. In other words, the industry grows and spreads into new markets because the public is eager to gamble. Furthermore, opinion polls show that most Americans support legal gambling opportunities, particularly lotteries and casinos.

Gambling is one of the most popular leisure activities in the country. According to the American Gaming Association (AGA), in *2016 State of the States: The AGA Survey of the Casino Industry* (November 2016, https://www.americangaming.org/sites/default/files/2016%20State%20of%20the%20States_FINAL.pdf), commercial casinos generated revenues of $38.5 billion in 2015. The Center for Gaming Research at the University of Nevada, Las Vegas, provides an even higher revenue total, estimating that casino gaming operations generated $40.2 billion that year. (See Table 2.1.) In 2016 tribal casinos took in $31.2 billion, according to the National Indian Gaming Commission in "Gross Gaming Revenue Trending" (July 16, 2017, https://www.nigc.gov/images/uploads/reports/2016GrossGamingRevenueTrends.pdf). These numbers dwarf, for example, domestic box office sales of the U.S. motion picture industry. The Motion Picture Association of America indicates in *Theatrical Market Statistics, 2016* (March 2017, http://www.mpaa.org/wp-content/uploads/2017/03/MPAA-Theatrical-Market-Statistics-2016_Final.pdf) that the U.S. movie industry earned $11.4 billion in box office sales in the U.S./Canadian market in 2016.

By contrast, gambling opponents argue that supply drives demand. They surmise that people would not be tempted to gamble or to gamble as often if opportunities were not so prevalent and widespread. They view gambling as an irresistible temptation with potentially dangerous consequences. It bothers them that gambling opportunities are presented, promoted, and supported not only by the business world but also by government leaders and politicians—people who are supposed to represent the best interests of the public they serve.Whatever the driving reason, gambling has become a big business and a popular pastime for many Americans, and since the 1980s the trend in public opinion and the legal system has been toward gambling expansion rather than prohibition.

SUPPLY: GAMBLING OPPORTUNITIES AND OPPORTUNISTS

A variety of gambling opportunities are available in the United States, both legal and illegal. Gambling is a moneymaking activity for corporations, small businesses, charities, governments, and, in some cases, criminals. The legal gambling industry employs hundreds of thousands of people across the country. In addition, it generates business in a variety of related industries, including manufacturing companies that provide slot machines and other supplies; travel and tourism companies that provide transportation, food, and lodging for gamblers; advertising agencies that promote gambling enterprises; and breeders who raise and train greyhounds and racehorses.

Gross gambling revenue is the money that is taken in by the industry minus the winnings paid out. In other words, it is equivalent to sales. From this number, then, operating expenses such as wages, benefits, and taxes must be subtracted to gauge the profits that are realized by the industry. In *2016 State of the States*, the AGA estimates that of the $38.5 billion in gross revenues earned by the commercial casino industry in 2015, $14.4 billion was paid out in wages (including benefits and tips) and $8.9 billion was paid in state and local taxes.

Casino Owners and Operators

Corporations have profited the most from legalized gambling since they took over the small casinos of Las Vegas, Nevada, during the late 1960s. The government had pushed out organized crime, which enabled the corporations

TABLE 2.1

Casino gaming revenues, by state, 2006–15

	2006	2007	2008	2009	2010	2011	2012	2013	2014	2015
Colorado	782,099	816,130	715,880	734,591	759,610	750,109	766,254	748,707	745,898	794,761
Delaware	651,734	612,407	588,923	564,239	571,376	547,872	549,643	471,544	408,206	400,854
Florida	10,300	201,132	225,290	216,747	329,127	381,122	427,889	467,587	507,454	530,662
Illinois	1,923,528	1,983,387	1,568,727	1,428,923	1,373,422	1,477,601	1,638,168	1,551,312	1,465,353	2,352,427
Indiana	2,576,192	2,623,939	2,665,663	2,798,195	2,796,005	2,732,773	2,685,503	2,434,284	2,156,766	2,132,889
Iowa	1,253,710	1,363,055	1,419,545	1,380,744	1,368,074	1,423,998	1,467,165	1,416,717	1,396,000	1,424,352
Kansas	—	—	—	1,990	37,788	44,729	341,146	365,079	353,539	367,783
Louisiana	2,567,415	2,566,271	2,583,834	2,455,526	2,373,930	2,374,244	2,388,767	2,442,900	2,472,502	3,242,009
Maine	37,517	43,252	50,515	59,198	61,667	59,453	82,994	126,274	540,483	516,615
Maryland	—	—	—	—	27,596	155,709	377,814	746,914	931,092	1,098,426
Michigan	1,303,303	1,335,016	1,359,585	1,339,479	1,377,929	1,424,445	1,416,734	1,349,504	1,332,783	1,376,408
Mississippi	2,570,884	2,891,546	2,721,139	2,464,662	2,388,997	2,239,084	2,251,090	2,136,624	2,070,157	2,097,066
Missouri	1,592,000	1,592,000	1,682,000	1,735,000	1,795,000	1,815,000	1,775,000	1,706,738	1,660,097	1,701,896
Nevada	12,622,044	12,849,137	11,599,124	10,392,675	10,404,731	10,700,994	10,860,715	11,142,915	11,009,684	11,114,081
New Jersey	5,217,613	4,920,786	4,544,961	3,943,171	3,565,047	3,317,720	3,051,435	2,862,069	2,742,128	2,536,729
New Mexico	238,310	244,780	258,080	243,940	247,350	248,920	241,480	241,300	257,660	265,960
New York	426,305	828,205	947,275	1,019,279	1,087,749	1,259,813	1,802,212	1,925,565	1,898,336	1,950,964
Ohio	—	—	—	—	—	—	429,826	1,070,662	1,457,634	1,642,903
Oklahoma	73,675	78,698	92,477	94,130	99,881	106,229	113,055	112,853	113,370	111,370
Pennsylvania	31,568	1,039,031	1,615,566	1,964,570	2,486,408	3,025,048	3,158,318	3,113,929	3,069,078	3,173,789
Rhode Island	406,504	447,998	475,040	461,169	477,050	512,865	527,959	516,742	550,912	551,923
South Dakota	89,828	98,223	102,264	101,898	106,187	100,898	107,384	103,019	100,509	101,680
West Virginia	975,990	932,210	951,210	905,590	877,650	958,700	948,810	796,589	704,947	722,170
Total	**35,350,519**	**37,467,203**	**36,167,098**	**34,305,716**	**34,612,574**	**35,657,326**	**37,409,361**	**37,849,827**	**37,944,585**	**40,207,717**

SOURCE: David G. Schwartz, "United States Annual Commercial Casino Gaming Revenues, 2006–2015," in *United States Commercial Gaming Revenues*, University of Nevada, Las Vegas, Center for Gaming Research, June 2016, http://gaming.unlv.edu/reports/national_annual_revenues.pdf (accessed June 26, 2017)

to bring their management practices to an increasingly profitable business. They invested money in new and bigger properties in Las Vegas and throughout the state of Nevada and then, in 1978, opened the first casino hotel in Atlantic City, New Jersey. As of 2017, most corporations in the industry owned or operated several commercial casinos. Casino City Press estimates in *North American Gaming Almanac* (2017) that in 2014 commercial casinos and card rooms accounted for 46% of the legal gambling revenues in the United States and Canada, tribal gaming accounted for 28%, lotteries accounted for 25%, and race and sports wagering accounted for 3%. Some companies, such as the Caesars Entertainment Corporation, also manage casinos for Native American tribes. The tribes are increasingly partnering with large, well-known corporations to take advantage of their name recognition and corporate experience.

Many of the nation's gambling properties are controlled by five well-known corporations: MGM Resorts International, Caesars Entertainment Corporation, Boyd Gaming Corporation, Penn National Gaming, and Red Rock Resorts.

MGM RESORTS INTERNATIONAL. MGM Resorts International reports in *2016 Annual Report* (April 2017, http://mgmresorts.investorroom.com/download/2016+MGM+Annual+Report.pdf) that it generated $9.5 billion in revenues in 2016, up from $9.2 billion in 2015. The company's domestic holdings included nine casinos in Las Vegas (Bellagio, MGM Grand Las Vegas, Mandalay Bay, The Mirage, Luxor, New York–New York, Excalibur, Monte Carlo, and Circus Circus Las Vegas), as well as casinos in Michigan, Mississippi, New Jersey, and Maryland. The company's other major segment was a 56% stake in MGM China, which consisted of MGM Macau and a second casino in the Cotai section of Macau that was still in development.

CAESARS ENTERTAINMENT CORPORATION. According to the Caesars Entertainment Corporation's 10-K filing (February 15, 2017, http://files.shareholder.com/downloads/ABEA-5FED0N/4979713972x0x931987/3B1625BD-3B09-4205-8FCF-7B7306425098/CZR_2016_Form_10-K.pdf) with the U.S. Securities and Exchange Commission (SEC), in 2016 the company owned or managed 12 casinos in the United States. Eight of the casinos were located in Las Vegas. In 2016 Caesars employed approximately 31,000 people and had net revenues of $3.9 billion.

BOYD GAMING CORPORATION. In its 10-K filing (February 2017, http://boydgaming.investorroom.com/download/Boyd+Gaming+2016+10-K.PDF) with the SEC, the Boyd Gaming Corporation explains that it owned 24 casinos in seven states in 2016. These included nine casinos in Las Vegas, among them the Fremont Hotel and Casino, the Gold Coast Hotel and Casino, and the Orleans Hotel and Casino, as well as casinos in Illinois, Indiana, Iowa, Kansas, Louisiana, and Mississippi. The corporation generated net revenues of $2.2 billion in 2016.

PENN NATIONAL GAMING. Penn National Gaming reports in its 10-K filing (February 2017, https://pennnationalgaming.gcs-web.com/static-files/2228b7f4-9519-4e77-a278-af88492fd184) with the SEC that in 2016 it owned, managed, or had ownership stakes in 27 gaming and racing facilities spread across Florida, Illinois, Indiana, Kansas, Maine, Massachusetts, Mississippi, Missouri, Nevada, New Jersey, New Mexico, Ohio, Pennsylvania, Texas, West Virginia, California, and Ontario, Canada. Penn National Gaming had net earnings of just over $3 billion in 2016, up from $2.8 billion in 2015. As of December 2016, it employed 18,808 people.

RED ROCK RESORTS. Red Rock Resorts indicates in its 10-K filing (March 2017, https://www.last10k.com/sec-filings/rrr/0001653653-17-000004.htm#fullReport) with the SEC that in 2016 it fully or partially owned and operated 20 casinos in the Las Vegas metropolitan area, while also managing tribal casinos in California and Michigan. Red Rock Resorts is a holding company that conducts business through the subsidiary Station Casinos LLC, which oversees casino operations. The company reported net revenues of $1.5 billion in 2016, up from $1.4 billion the year before. As of January 2017, the company employed approximately 13,650 people in the properties in which it had a full or majority ownership share.

Other Gambling Corporations

Besides numerous smaller casino and gaming corporations, the gambling industry supports many related companies, such as those that provide equipment, goods, supplies, and services. According to the Association of Gaming Equipment Manufacturers (2016, http://www.agem.org/), in 2016 the gaming equipment manufacturing segment accounted for $17.9 billion in direct economic output and employed 55,145 workers who collectively earned $4.9 billion in wages. Among the many companies that supply casinos and other gaming operators are:

- Bally Technologies, Inc., which introduced its first slot machine in 1936 and is now a successful machine manufacturer and distributor
- International Game Technology PLC, which introduced the first lottery terminal in 1982 and now provides technology services to lotteries in several U.S. states
- Konami Gaming, Inc., a Japanese producer of high-tech video slot machines and multisite casino management systems that entered the U.S. gaming market in 2000
- WMS Gaming, which is engaged entirely in the manufacture, sale, leasing, and licensing of gambling machines

Small Businesses

Many small casinos and racetracks, minicasinos, and card rooms around the country are owned or operated by small companies, families, and entrepreneurs. Other ways in which small businesses are engaged in or serve the gambling industry include:

- Selling lottery tickets and operating electronic gaming devices at independently owned convenience stores, markets, service stations, bars, restaurants, bowling alleys, and newsstands
- Manufacturing and distributing equipment such as slot machines, roulette wheels, lottery tickets, dice, and cards
- Providing services such as advertising, marketing, public relations, accounting, information technology, and food
- Breeding, training, and caring for horses and greyhounds

Most small businesses that offer gambling do so through lottery ticket sales or electronic gaming devices, such as slot machines. These are considered forms of convenience gambling because patrons do not have to travel to special destinations, such as casinos and racetracks. Convenience gambling has been more controversial than destination gambling. Critics argue that allowing gambling in stores, restaurants, and other places that people visit as part of their everyday routine makes it too easy for them to gamble. The same criticism is leveled against Internet gambling, which patrons can do at home.

Internet Gambling Businesses

Between 1995 and 2006 Internet gambling in the United States was largely the province of small companies operating offshore, many of them located in the Caribbean or Europe. During this period Internet gambling operated in a legal gray area and was unregulated. Most online gambling operators paid little or no taxes, and little was known about their profitability. Online poker was a particularly popular activity, thanks in large part to the dual trends of increasing Internet access and the growing popularity of the World Series of Poker, a televised event that pits expert poker players against one another. Online poker sites began serving as forums for players hoping to qualify for the televised event, and the field of contestants exploded between 2003 and 2006.

In 2006 Congress passed the Unlawful Internet Gambling Enforcement Act (UIGEA), which made it illegal for banks and credit card companies to transfer money to operators of gambling websites. Most casual online bettors in the United States stopped gambling, and many website operators withdrew from the U.S. market. Serious players, however, continued to play online. In 2011 the U.S. Department of Justice cracked down on three online poker companies (Full Tilt Poker, Absolute Poker, and PokerStars) that continued to allow U.S. residents to gamble online by seizing the companies' assets and website domain names.

Later that same year, however, the Department of Justice issued the opinion *Whether Proposals by Illinois and New York to Use the Internet and Out-Of-State Transaction Processors to Sell Lottery Tickets to In-State Adults Violate the Wire Act* (September 20, 2011, https://www.justice.gov/sites/default/files/olc/opinions/2011/09/31/state-lotteries-opinion_0.pdf), which indicated that states had the leeway to legalize and regulate Internet gambling, with the exception of online sports betting. Delaware, Nevada, and New Jersey responded by legalizing online gambling. The three states structured their online gambling systems differently. In Delaware gambling websites are operated by the state lottery, whereas in Nevada and New Jersey the websites are operated by existing casino operators. Delaware offers lottery sales and a variety of casino games, New Jersey offers all games that are available in offline casinos, and Nevada offers only poker.

As of 2017, online sports betting remained illegal in all U.S. jurisdictions. Those companies that operated Internet sports gambling websites did so in violation of federal law.

Criminals

Gambling has had a checkered legal history in the United States. At various times it has been legal, illegal but tolerated, or illegal and actively prosecuted. During times when gambling opportunities have been outlawed, entrepreneurs have stepped in to offer them anyway. These entrepreneurs range from mobsters running million-dollar betting rings to grandmothers running neighborhood bingo games. In any case, the illegal nature of the activity makes such entrepreneurs criminals.

Organized crime groups have often been associated with gambling, which is a cash business with high demand and good profits. Crime syndicates on the East Coast were among the first to see the potential of Las Vegas, invest in it, and profit from it. At times, they infiltrated other segments of the legal gambling industry, such as horse racing. Nevertheless, strict regulations and crackdowns by law enforcement pushed them out as the gambling industry modernized. The federal Racketeer Influenced and Corrupt Organizations Act of 1970 was designed to combat infiltration by organized crime into legitimate businesses, including gambling enterprises. Most analysts believe this act has been largely successful at keeping mobsters from establishing or taking over legal gambling businesses.

Although they have been denied casino ownership and management roles, some organized crime figures have infiltrated casinos in other ways, such as through labor unions and maintenance or food services. Law enforcement officials also believe organized crime families have been involved in bribing state officials who were considering the extension or expansion of gambling options, particularly relating to electronic gambling machines.

The Nevada Gaming Commission and the State Gaming Control Board maintain a list of people who are prohibited from gambling in Nevada. The List of Excluded Persons, more commonly known as "Nevada's Black Book," includes known cheaters, crime family bosses, mob associates, and others who are linked in some way to organized crime. These people are considered so dangerous to the integrity of legal gambling that they are not allowed to set foot in Nevada casinos. In "GCB Excluded Person List" (May 2017, http://gaming.nv.gov/index.aspx?page=72), the Nevada Gaming Commission and the State Gaming Control Board provide photographs of these excluded people.

The most lucrative sector of the gambling industry for organized crime has been and continues to be illegal bookmaking and numbers games. Bookmaking is a gambling activity in which a bookmaker takes bets on the odds that a particular event will occur or that an event will have a particular outcome. The vast majority of bookmaking revolves around sporting events, such as college and professional football and basketball games. Such wagering is extremely popular in the United States. Because sports bookmaking is legal only in Nevada and, in a very limited form, in Delaware, there is a large illegal market for it across the country. According to the Spectrum Gaming Group, in *Gambling Impact Study* (October 28, 2013, http://www.leg.state.fl.us/GamingStudy/docs/FGIS_Spectrum_28Oct2013.pdf), the state of Nevada reported that $3.4 billion was wagered in legal sports betting in 2012; by comparison, illegal sports wagers total an estimated $80 billion to $380 billion annually.

Illegal numbers games are similar to lottery games in that players wager money on particular numbers to be selected in a drawing or by other means. Illegal numbers operators thrive in many parts of the country, especially large cities—even those where legal lotteries are offered.

Not all bookmaking is done through mobsters. Many enterprising entrepreneurs run small-time illegal gambling books, mostly related to sporting events. Office pools, in which coworkers pool small wagers on sports or office events (e.g., when a baby is going to be born) are common. Although society does not generally consider private wagers and small-stakes office pools to be illegal gambling, the laws in most states do.

Despite widespread illegal gambling, few people are actually arrested for engaging in it. Authorities made only 4,825 arrests for gambling in 2015 out of 10.8 million total arrests that year. (See Table 2.2.)

Charities

Charitable gambling is the most widely practiced form of gambling in the United States. In 2017 it was legal in 48 states and the District of Columbia (only Hawaii and Utah prohibited it). In charitable gambling, a specified portion of the money that is raised (minus

TABLE 2.2

Estimated number of arrests by type of crime, 2015

Total[a]	**10,797,088**
Murder and nonnegligent manslaughter	11,092
Rape[b]	22,863
Robbery	95,572
Aggravated assault	376,154
Burglary	216,010
Larceny-theft	1,160,390
Motor vehicle theft	77,979
Arson	8,834
Violent crime[c]	505,681
Property crime[c]	1,463,213
Other assaults	1,081,019
Forgery and counterfeiting	55,333
Fraud	133,138
Embezzlement	15,909
Stolen property; buying, receiving, possessing	88,576
Vandalism	191,015
Weapons; carrying, possessing, etc.	145,358
Prostitution and commercialized vice	41,877
Sex offenses (except rape and prostitution)	51,388
Drug abuse violations	1,488,707
Gambling	4,825
Offenses against the family and children	94,837
Driving under the influence	1,089,171
Liquor laws	266,250
Drunkenness	405,880
Disorderly conduct	386,078
Vagrancy	25,151
All other offenses	3,218,880
Suspicion	1,389
Curfew and loitering law violations	44,802

[a]Does not include suspicion.
[b]The rape figure in this table is an aggregate total of the data submitted based on both the legacy and revised Uniform Crime Reporting definitions.
[c]Violent crimes are offenses of murder and nonnegligent manslaughter, rape, robbery, and aggravated assault. Property crimes are offenses of burglary, larceny-theft, motor vehicle theft, and arson.

SOURCE: "Table 29. Estimated Number of Arrests: United States, 2015," in *Crime in the United States, 2015*, U.S. Department of Justice, Federal Bureau of Investigation, 2016, https://ucr.fbi.gov/crime-in-the-u.s/2015/crime-in-the-u.s.-2015/tables/table-29 (accessed June 26, 2017)

prizes, expenses, and any state fees and taxes) goes to qualified charitable organizations. Such organizations include religious groups, fraternal organizations, veterans groups, volunteer fire departments, parent-teacher organizations, civic and cultural groups, booster clubs, and other nonprofit organizations.

Generally, a charitable organization has to have been in existence for several years and has to obtain a state license for the gambling activity. Most states will only issue licenses to organizations that have been recognized by the Internal Revenue Service as exempt from federal income tax under Tax Code section 501(c). Thousands of charitable organizations are registered to conduct gambling throughout the country.

Most charitable gambling is regulated by state governments, although not uniformly by the same department—it may be the department of revenue, the state police, the alcohol control board, or the lottery, gaming, or racing commission. Administrative fees and taxes are levied in most states. In some states charitable gambling activity is unregulated.

Typical games allowed include bingo (the most common), pull tabs (lottery tickets with tabs that gamblers pull open to reveal cash prizes), raffles, and card games such as poker or blackjack. Slot machines and table games such as roulette and craps are generally not permitted. Limits are usually placed on the size of cash prizes that can be awarded. The games that are allowable by law for charity fund-raising vary from state to state.

Because of the inconsistencies in state oversight, it is difficult to determine the complete extent of charitable gambling in the United States. In *NAFTM 2015 Annual Report* (2016, http://www.naftm.org/vertical/sites/%7B10B16680-A509-4D78-B468-8A1901FC0CF7%7D/uploads/NAFTM15_FINAL.pdf), the National Association of Fundraising Ticket Manufacturers, a trade association representing companies that manufacture bingo paper, pull tabs, and other supplies used in the charitable gambling industry, provides the best available estimates for the scope of charitable gaming in the United States. In terms of gross receipts, the leading states for charitable gaming in 2015 were Minnesota ($1.4 billion), Texas ($756.8 million), Washington ($442.7 million), Indiana ($394.3 million), and Kentucky ($346.6 million). Washington was the leading charitable gaming state in terms of net proceeds, with $289.8 million, followed by Minnesota ($254.2 million), Ohio ($92 million), Indiana ($61.2 million), Kentucky ($41.3 million), and New York ($40.5 million). Pull tabs were the largest sources of charitable gaming revenues in most states, followed by bingo and raffles.

The Government

Federal, state, local, and tribal agencies collect money from gambling operations through the assessment of taxes and fees and, in some cases, by directly supplying gambling opportunities. The expansion of gambling often seems an attractive solution to budget deficits and economic stagnation, and many of the laws passed to legalize gambling have been couched in these terms. One example is Nevada, which legalized gambling in 1931, at the height of the Great Depression (1929–1939). Another is New Jersey, which legalized gambling in Atlantic City in 1976 largely as a means of providing an economic boost to the troubled area. Similarly, the rapid expansion of legal gambling during the 1990s and the first decade of the 21st century overlapped substantially with recessionary economic environments, with proponents of legalization framing it as a way of generating tax revenues to fund schools and other public priorities. Moreover, on native tribal lands, which are often located in remote areas not conducive to other types of economic development, gaming revenues are in many cases an instrumental source of employment and tax revenues.

Despite these potential economic advantages, it remains unclear whether legalized gambling provides significant financial advantages to the community as a whole. Yixin Liu et al. of the Edward J. Bloustein School of Planning and Public Policy report in *Promises Made, Promises Broken: An Overview of Gambling in New Jersey and Recommendations for the Future* (May 2015, http://bloustein.rutgers.edu/wp-content/uploads/2015/08/Final-Gambling-Report.pdf) that although revenues derived from gambling operations in Atlantic City have generated enormous profits to casino owners, these funds are rarely reinvested into the community. In 2015 Atlantic City had an unemployment rate of 18.4%, the highest in the state, and a poverty rate of 34%. Furthermore, as Liu et al. note, the casino industry has been shown to exacerbate social problems such as gambling addiction and crime, issues that disproportionately affect low-income citizens.

THE FEDERAL GOVERNMENT. The primary means by which the federal government makes money from the gambling industry is by taxing winning gamblers and gambling operators. Gamblers must declare gambling earnings when they file their personal income taxes. They get to subtract their gambling losses, but they must keep thorough records and have receipts, if possible, to prove their losses. For racetrack gamblers, this means saving losing betting slips and keeping a gambling diary of dates, events, and amounts. Casino gamblers who join so-called slot clubs can get a detailed printout of their gambling history from the casino.

Gambling operators, like all companies, are subject to corporate taxes. They are required to report winnings that meet certain criteria to the Internal Revenue Service. (See Table 2.3.) The gambling operator must withhold income tax from winnings of more than $5,000, usually at a withholding rate of 25%. Gamblers who win noncash prizes, such as cars or other merchandise, have to pay taxes on the fair market value of the items.

STATE GOVERNMENTS. State governments benefit significantly from tax revenues generated by commercial casinos and lotteries, and they also derive tax revenues from pari-mutuel and other forms of gambling. According to Michael Pollock of the Spectrum Gaming Group, in *Casino Tax Policy: Identifying the Issues That Will Determine the Optimal Rate* (2010, https://www.spectrumgaming.com/dl/SpectrumNationalTaxAssociation.pdf), Nevada's casinos are taxed at an effective rate of 6.8% and New Jersey's are taxed at an effective rate of 8.4%. The states that have legalized commercial casinos since the 1990s have tended to impose much higher effective tax rates on gambling revenues. Except for Mississippi (which has an effective tax rate of 12%), all of these states have imposed effective tax rates of more than 20% on legal gaming operators, and a number of them have rates that approach or exceed 50%, including Maine (49.1%), Florida (50%), Illinois (50%), Pennsylvania (55%), West Virginia (56.7%), Delaware (56.9%), New York (65%), and Rhode Island (72.7%).

Thus, even though Nevada's gaming industry dwarfs that of all other states, it trails behind Pennsylvania in terms of the amount of tax revenues that are generated for state and local governments. As Table 2.4 reveals, in 2015 consumers spent $11.1 billion on gambling in Nevada, compared with $3.2 billion on gambling in Pennsylvania. Casinos in Nevada, however, paid $889.1 million in state and local taxes that year, whereas casinos in Pennsylvania paid $1.4 billion in taxes. (See Table 2.5.) Among the other

TABLE 2.3

Gambling winnings that must be reported to the Internal Revenue Service, 2017

Type of game	Amount of winnings exceeds	Reduced by wager?
Lottery, sweepstakes, wagering pool, horse race, dog race, jai alai, or other wagering transactions	$600 and 300 times amount of wager	Yes
Bingo or slot machine	$1,200	No
Keno	$1,500	Yes
Poker tournament	$5,000	Yes

SOURCE: Adapted from "2017 Instructions for Forms W-2G and 5754," U.S. Department of the Treasury, Internal Revenue Service, 2017, https://www.irs.gov/pub/irs-pdf/iw2g.pdf (accessed June 26, 2017)

TABLE 2.4

Consumer spending on commercial casino gaming, by state, 2014 and 2015

State	2014	2015	% change
Colorado	$746.35M	$790.08M	5.86%
Delaware	$411.71M	$410.49M	−0.30%
Florida	$507.47M	$530.66M	4.57%
Illinois	$ 1.463B	$ 1.438B	−1.73%
Indiana	$ 2.232B	$ 2.216B	−0.76%
Iowa	$ 1.396B	$ 1.424B	2.03%
Kansas	$353.54M	$367.77M	4.03%
Louisiana	$ 2.472B	$ 2.649B	7.12%
Maine	$127.27M	$129.81M	1.99%
Maryland	$931.08M	$ 1.098B	17.97%
Massachusetts (opened June 2015)	—	$ 88.23M	—
Michigan	$ 1.333B	$ 1.376B	3.27%
Mississippi	$ 2.070B	$ 2.097B	1.30%
Missouri	$ 1.660B	$ 1.702B	2.52%
Nevada	$ 11.019B	$ 11.114B	0.87%
New Jersey	$ 2.742B	$ 2.563B	−6.53%
New Mexico	$ 257.66B	$256.02M	−0.64%
New York	$ 1.898B	$ 1.951B	2.77%
Ohio	$ 1.458B	$ 1.644B	12.77%
Oklahoma	$111.37M	$113.14M	1.59%
Pennsylvania	$ 3.069B	$ 3.174B	3.41%
Rhode Island	$611.08M	$615.82M	0.78%
South Dakota	$104.06M	$108.36M	4.13%
West Virginia	$722.17M	$686.56M	−4.93%
Total	**$37.696B**	**$38.543B**	**2.25%**

M = million
B = billion

SOURCE: "State-by-State Consumer Spending on Commercial Gaming, 2014 vs. 2015," in *2016 State of the States: The AGA Survey of the Casino Industry*, American Gaming Association, November 15, 2016, https://www.americangaming.org/sites/default/files/2016%20State%20of%20the%20States_FINAL.pdf (accessed June 26, 2017)

TABLE 2.5

Tax revenues generated by casino gaming, by state, 2014 and 2015

State	2014	2015	% change
Colorado	$106.13M	$113.80M	7.23%
Delaware	$168.32M	$160.96M	−4.38%
Florida	$177.61M	$185.73M	4.57%
Illinois	$ 500.6M	$488.04M	−2.51%
Indiana	$622.14M	$608.06M	−2.26%
Iowa	$316.37M	$329.35M	4.10%
Kansas	$ 88.38M	$ 92.01M	4.10%
Louisiana	$591.25M	$632.24M	6.93%
Maine	$ 51.35M	$ 52.21M	1.67%
Maryland	$415.91M	$452.92M	8.90%
Massachusetts (opened June 2015)	—	$ 43.23M	—
Michigan	$302.95M	$311.24M	2.74%
Mississippi	$246.56M	$252.41M	2.37%
Missouri	$436.94M	$443.89M	1.59%
Nevada	$ 881.5M	$889.13M	0.87%
New Jersey	$253.65M	$237.09M	−6.53%
New Mexico	$ 67.19M	$ 67.15M	−0.06%
New York	$864.18M	$888.37M	2.80%
Ohio	$484.65M	$545.38M	12.53%
Oklahoma	$ 20.63M	$ 20.76M	0.60%
Pennsylvania	$ 1.348B	$ 1.379B	2.34%
Rhode Island	$333.52M	$328.84M	−1.40%
South Dakota	$ 15.86M	$ 15.67M	−1.20%
West Virginia	$314.75M	$316.45M	0.54%
Total	**$ 8.608B**	**$ 8.854B**	**2.86%**

M = million
B = billion

SOURCE: "Commercial Casino Tax Revenue by State, 2014 vs. 2015," in *2016 State of the States: The AGA Survey of the Casino Industry*, American Gaming Association, November 15, 2016, https://www.americangaming.org/sites/default/files/2016%20State%20of%20the%20States_FINAL.pdf (accessed June 26, 2017)

states that benefited most from taxes paid by casinos, Pollock indicates there were a mix of effective rates, ranging from New York (65%) and Rhode Island (72.7%) to Louisiana (21.5%), Missouri (21%), and Iowa (23.2%). The optimal rate at which taxes for casinos should be set to create the most benefit for state and local governments remains a subject of debate.

Unlike casinos, which fund government services through the taxes they pay, lotteries are directly operated by states. As of 2017, 44 states and the District of Columbia operated lotteries. The only states that did not operate lotteries were Alabama, Alaska, Hawaii, Mississippi, Nevada, and Utah. According to the North American Association of State and Provincial Lotteries (2017, http://www.naspl.org/wherethemoneygoes/), state lotteries contributed approximately $22.6 billion to state and local government programs in 2016.

In Kansas commercial casinos are owned by the state. As of 2017, there were four such casinos in the state. In Delaware the state lottery operates the state's nascent (developing) online gambling websites. Also in Delaware, as well as in a number of other states (New York, Ohio, Rhode Island, and West Virginia), privately owned casinos offer video lottery terminals that are operated by the state.

TRIBAL GOVERNMENTS. Native American tribes that operate casinos and other gaming businesses use gambling revenues to fund tribal government operations or programs, to provide for the general welfare of the tribe and its members, to promote tribal economic development, to make charitable contributions, and to fund local government agencies, among other uses. According to the National Indian Gaming Commission, in the press release "2016 Indian Gaming Revenues Increased 4.4%" (July 17, 2017, https://www.nigc.gov/news/detail/2016-indian-gaming-revenues-increased-4.4), 244 tribes in 28 U.S. states were engaged in gaming operations in 2016. Collectively, these tribes generated $31.2 billion in revenues that year.

DEMAND: THE GAMBLERS

Gambling is a leisure activity. People gamble because they enjoy it. Proponents say there is no difference between spending money at a theme park and spending it at a casino: the money is exchanged for a good time in either case. Nevertheless, gambling has a powerful allure besides fun: the dream of wealth, which is a strong motivator. Some options, such as lotteries, offer the chance to risk a small investment for an enormous payoff. This potential is too appealing for many people to pass up.

Adults

In "March Madness and American Gambling Habits" (March 15, 2012, http://www.marketplace.org/topics/life/march-madness-and-american-gambling-habits), Jeremy Hobson interviews the *Los Angeles Times* columnist David Lazarus, who states that 86% of Americans participate in some form of gambling at least once per year and that 16% of Americans gamble on a weekly basis. Overall, Americans who gamble lose roughly $100 billion annually.

Paul Taylor, Cary Funk, and Peyton Craighill of the Pew Research Center note in *Gambling: As the Take Rises, So Does Public Concern* (May 23, 2006, http://assets.pewresearch.org/wp-content/uploads/sites/3/2010/10/Gambling.pdf), the most recent authoritative overview of the U.S. gambling population as of August 2017, that approximately two-thirds (67%) of Americans gambled between March 2005 and March 2006. Men (72%) were more likely than women (62%) to have gambled during this period. Those with a college education were slightly less likely to have gambled than were those with only some college or with a high school diploma or less. The participation rate for people with higher incomes (greater than $100,000 per year) was higher than for other income groups. Only

59% of those making less than $30,000 per year reported gambling between March 2005 and March 2006, compared with 79% of those making more than $100,000 per year.

The AGA offers in *2013 State of the States: The AGA Survey of Casino Entertainment* (2013, https://www.americangaming.org/sites/default/files/research_files/aga_sos2013_rev042014.pdf) a more recent profile of the gambling population that is roughly consistent with the overview provided by Taylor, Funk, and Craighill. The AGA finds that the most common form of gambling participation is the purchase of lottery tickets. In 2012, 53% of AGA survey respondents played the lottery, 32% gambled in a casino, 26% engaged in casual wagers with friends, 12% played poker, 6% wagered on a race, and 3% gambled on the Internet. Also in line with Taylor, Funk, and Craighill, the AGA notes that the gambling population is slightly wealthier, on average, than the U.S. population at large. Approximately half (49%) of all casino visitors in 2012 reported household incomes in excess of $60,000 per year, whereas only 34% of all surveyed households reported more than $60,000 in income. In contrast with Taylor, Funk, and Craighill, the AGA indicates that casino visitors in 2012 were slightly more educated, on average, than all people surveyed by the AGA. More than half (52%) of casino gamblers reported having a college degree, whereas only 46% of the nationally representative survey sample had completed college.

In *Gambling Impact Study*, the Spectrum Gaming Group, noting a lack of recent authoritative data, provides an estimate of gambling prevalence rates based on a survey of past studies on the topic: "Various studies commissioned by individual states since 1976 have shown lifetime prevalence rates ranging from 64 percent to 96 percent, with past-12-month prevalence rates [the rates of gambling activity in the preceding 12 months] ranging even more broadly from between 49 percent to 89 percent. A meta-analysis of available research across the United States and Canada conducted in 1997 estimated a lifetime gambling prevalence rate of 81 percent in the general population across the country as a whole."

SENIOR CITIZENS. Although anecdotal evidence suggests that the elderly are disproportionately likely to gamble, especially in commercial casinos, the data on the subject are inconclusive. According to Taylor, Funk, and Craighill, 58% of people aged 65 years and older reported gambling between March 2005 and March 2006. Other studies show similar, or even higher, rates of participation. For example, Suzi Levens et al. reveal in "Gambling among Older, Primary-Care Patients" (*American Journal of Geriatric Psychiatry*, vol. 13, no. 1, January 2005) that in 2005 nearly 70% of Americans older than 65 years reported gambling in the previous 12 months. This study, which was based on a survey of 843 elderly patients, also finds that nearly 11% of those questioned were at risk for problem gambling. In contrast, the AGA maintains in *2013 State of the States* that the elderly gamble in casinos at lower rates than younger adults. In 2012, 28% of adults aged 65 years and older visited a casino, compared with 39% of those aged 21 to 35 years, 34% of those aged 36 to 49 years, and 36% of those aged 50 to 64 years.

For older adults not at risk for gambling problems, the activity may have a positive impact. In "Health Correlates of Recreational Gambling in Older Adults" (*American Journal of Psychiatry*, vol. 161, no. 9, September 2004), Rani A. Desai et al. indicate that a correlation exists between gambling and good health among people older than the age of 65. They note such health benefits include opportunities for socializing among peers, as well as the stimulation of cognitive and sensory faculties. Desai et al. do not find a similar correlation in those aged 18 to 64 years. The study, which is based on interviews with 2,417 older adults, focuses only on recreational gamers and does not include subjects who exhibited gambling addiction.

Young People

The minimum legal age for placing a legal bet ranges from 18 to 21 years, depending on the state and the activity. For example, most states limit the sale of lottery tickets to those who are 18 years and older, although most allow minors to receive lottery tickets as gifts. All commercial casinos have a minimum gambling age of 21 years as set by state law. Tribal casinos are allowed to set their own minimum gambling age as long as it is at least 18 years. The minimum age to participate in charitable gambling activities, such as bingo games, is 18 years in most states. A few states allow people as young as 16 years to participate.

Social scientists consider emerging adults (people aged 18 to 25 years) to be at risk for developing a range of addictions. In "Examining Gender Differences for Gambling Engagement and Gambling Problems among Emerging Adults" (*Journal of Gambling Studies*, vol. 29, no. 2, June 2013), Gloria Wong et al. observe that emerging adults are more likely than adolescents or other adults to engage in gambling and to develop gambling problems. Although few comprehensive studies of this age group's gambling involvement have been conducted, Wong et al. point to surveys estimating that 42% of emerging adults engage in gambling and that 7% are problem gamblers. These estimates are broadly consistent with the AGA's finding in *2013 State of the States*, that adults aged 21 to 35 years gamble more frequently than any other age-based subgroup.

Since 2002 the Annenberg Public Policy Center at the University of Pennsylvania has been studying the gambling habits of young people as part of the National Annenberg Survey of Youth. The center notes in "Internet Gambling Grows among Male Youth Ages 18 to 22" (October 14, 2010, http://www.annenbergpublicpolicycenter.org/wp-content/uploads/Card-Playing-2010-Release-final.pdf) that 33.3% of young men aged 18 to 22 years played cards for money at least once per month in 2010, up slightly from 31.7% who gambled on cards monthly in 2008. In contrast, the percentage of young men between the ages of 18 and 22 years who gambled on the Internet once per month nearly quadrupled during this same span, from 4.4% in 2008 to 16% in 2010. According to Dan Romer, the director of the Annenberg Adolescent Communication Institute, roughly 1.7 million college-age males gambled online once per month in 2010; more than 400,000 gambled on the Internet once per week. Although the center recorded that no young women between the ages of 18 and 22 years gambled on the Internet in 2008, by 2010, 4.4% of females in that age group reported gambling online at least once per month. The most striking increase in gambling occurred among high school girls between the ages of 14 and 17 years. The center indicates that 9.5% of females in this age group gambled on sports once per month or more in 2008; by 2010 this proportion had grown to 22%. Overall, 28.2% of girls aged 14 to 17 years participated in some form of gambling at least once per month in 2010.

Problem Gamblers

Problem gambling is a broad term that covers all gambling behaviors that are harmful to people in some way—financially, emotionally, socially, and/or legally. The harmful effects of problem gambling include:

- Financial difficulties, such as unpaid bills, loss of employment, large debts, and even bankruptcy
- Emotional problems, such as depression, anxiety, addictions, and thoughts of suicide
- Social problems, as evidenced by strained or broken relationships with spouses, family, friends, and co-workers
- Legal problems related to neglect of children or commission of criminal acts to obtain money

Taylor, Funk, and Craighill indicate that in 2006, 6% of those asked said gambling had been a source of problems within their family. This percentage was up slightly from those reported by Gallup, Inc., in 1989, 1992, and 1996, but down slightly from the percentage reported in 1999. The researchers note that there is a marked difference in answers by age. Only 5% of those aged 50 years and older said gambling had been a problem for their family, compared with 12% of adults younger than 50 years.

In general, scientists characterize gambling behavior by the level of harm that it causes. People who experience no harmful effects are called nonproblem gamblers, or social, casual, or recreational gamblers. Those who gamble regularly and may be prone to a gambling problem are called at-risk gamblers, and those who experience minor to moderate harm from their gambling behavior are called problem gamblers. Pathological gamblers, as they have historically been known, are severely harmed by their gambling activities. As is discussed in Chapter 1, in the 2013 revision of the *Diagnostic and Statistical Manual of Mental Disorders* (*DSM-5*), the American Psychiatric Association (APA) reclassified its pathological gambling diagnosis as a gambling disorder, in the interest of removing the stigma associated with the former term. *Pathological gambling*, however, remains a term in common usage.

Scientists use a screening process to determine which category fits a particular gambler. One of the most common is the South Oaks Gambling Screen (SOGS), a 16-item questionnaire that was developed during the 1980s by Henry R. Lesieur and Sheila B. Blume. A detailed description of the questionnaire and its development was first presented by Lesieur and Blume in "The South Oaks Gambling Screen (SOGS): A New Instrument for the Identification of Pathological Gamblers" (*American Journal of Psychiatry*, vol. 144, no. 9, September 1987). The researchers used information from 1,616 subjects to develop SOGS, including patients with substance abuse and pathological gambling problems, Gamblers Anonymous members, university students, and hospital employees. Because potential problem gamblers fill out the questionnaire themselves, scores depend entirely on the truthfulness of the people answering the questions. Jeffrey N. Weatherly et al. indicate in "Validating the Gambling Functional Assessment—Revised in a United Kingdom Sample" (*Journal of Gambling Studies*, vol. 30, no. 2, June 2014) that as of 2014 SOGS continued to be the most widely used instrument to identify problem and pathological gamblers. The researchers suggest that once probable pathological gambling has been identified, a 20-item test should be used to identify what about gambling has made it addictive: to get attention from others, to experience the sensory stimulation, to escape from stress or personal problems, or to feel the excitement it brings. Once this assessment is completed, an appropriate treatment can be devised.

Another means of defining problem gamblers was created by the self-help organization Gamblers Anonymous, which prefers the term *compulsive gambling*. In "Questions & Answers about Gamblers Anonymous" (2017, http://www.gamblersanonymous.org/ga/content/questions-answers-about-gamblers-anonymous), the organization explains that compulsive gamblers exhibit certain characteristic behaviors, such as:

- An "inability and unwillingness to accept reality"
- Feelings of emotional insecurity when they are not gambling

- Immaturity and a desire to escape from responsibility
- Wanting all the good things in life without expending much effort for them
- Wanting to be a "big shot" in the eyes of other people

Gamblers Anonymous provides a list of 20 questions that gamblers can use to determine if they have a gambling problem. (See Table 2.6.) The organization indicates that compulsive gamblers are likely to answer yes to at least seven of the questions.

GAMBLING DISORDER. People with a gambling disorder, known historically as pathological gambling, are characterized as having irrational thoughts, in which they continuously (or periodically) lose control over their gambling behavior. Pathological gamblers become preoccupied with gambling, constantly thinking about their next bet or how to raise more money with which they can gamble. This behavior continues even if the gambler suffers adverse consequences, such as financial difficulties or strained relationships with family and friends.

Pathological Gambling: A Critical Review was published in 1999 by the National Academies Press. The book identifies and analyzes all available scientific research studies that deal with pathological and problem gambling. The studies are reviewed by dozens of researchers on behalf of the National Research Council, an organization that is administered by the National Academy of Sciences, the National Academy of Engineering, and the Institute of Medicine. The researchers estimate that 1.5% of U.S. adults have been pathological gamblers at some point during their life. In any given year 0.9% of U.S. adults (approximately 1.8 million people) and 1.1 million adolescents aged 12 to 18 years are pathological gamblers. The following general conclusions are drawn:

- Men are more likely than women to be pathological gamblers.
- Pathological gambling often occurs concurrently with other behavioral problems, such as drug and alcohol abuse and mood and personality disorders.
- The earlier in life a person starts to gamble, the more likely he or she is to become a pathological gambler.
- Pathological gamblers are more likely than those without a gambling problem to have parents who are pathological gamblers.
- Pathological gamblers who seek treatment generally get better.

The researchers, however, are unable to determine from available studies whether any particular treatment technique is more effective than most others or even if some pathological gamblers are able to recover on their own. They are also unable to determine whether particular groups, such as the elderly and the poor, have disproportionately high rates of pathological gambling. The researchers conclude that further studies are needed to provide a detailed understanding of pathological gambling.

In *DSM-5*, the APA reclassified pathological gambling as a gambling disorder and moved the disorder from the Impulse-Control Disorders Not Elsewhere Classified category to the Substance-Related and Addictive Disorders category. Nancy M. Petry et al. note in "An Overview of and Rationale for Changes Proposed for Pathological Gambling in DSM-5" (*Journal of Gambling Studies*, vol. 30, no. 2, June 2014) that this change in category reflects a number of research findings since 1994, when *DSM-IV* was published:

> Gambling and substance use disorders share similar presentations of some symptoms (Petry 2006; Toce-Gerstein et al. 2003), and the two disorders consistently demonstrate high rates of comorbidity in epidemiological as well as clinical samples (Kessler et al. 2008; Lorains et al. 2011; Nalpas et al. 2011; Petry et al. 2005b). Data are emerging that gambling and substance use disorders have common underlying genetic vulnerabilities (Black et al. 2006; Blanco et al. 2012; Slutske et al. 2000;), and both are associated with similar biological markers and cognitive deficits (Blanco et al. 2012; Potenza et al. 2003; Reuter et al. 2005). Effective treatments for gambling have been based on those for

TABLE 2.6

Twenty questions designed to determine whether a person is a compulsive gambler

1. Did you ever lose time from work or school due to gambling?
2. Has gambling ever made your home life unhappy?
3. Did gambling affect your reputation?
4. Have you ever felt remorse after gambling?
5. Did you ever gamble to get money with which to pay debts or otherwise solve financial difficulties?
6. Did gambling cause a decrease in your ambition or efficiency?
7. After losing did you feel you must return as soon as possible and win back your losses?
8. After a win did you have a strong urge to return and win more?
9. Did you often gamble until your last dollar was gone?
10. Did you ever borrow to finance your gambling?
11. Have you ever sold anything to finance gambling?
12. Were you reluctant to use "gambling money" for normal expenditures?
13. Did gambling make you careless of the welfare of yourself or your family?
14. Did you ever gamble longer than you had planned?
15. Have you ever gambled to escape worry, trouble, boredom or loneliness?
16. Have you ever committed, or considered committing, an illegal act to finance gambling?
17. Did gambling cause you to have difficulty in sleeping?
18. Do arguments, disappointments or frustrations create within you an urge to gamble?
19. Did you ever have an urge to celebrate any good fortune by a few hours of gambling?
20. Have you ever considered self-destruction or suicide as a result of your gambling?

SOURCE: "20 Questions: Are You a Compulsive Gambler?" Gamblers Anonymous, 2017, http://www.gamblersanonymous.org/ga/content/20-questions (accessed June 26, 2017)

substance use disorder (Grant et al. 2008; Hodgins et al. 2009; Petry et al. 2006, 2008). Gambling disorder also appears to align more closely to substance use disorders than to other psychiatric disorders (Blanco et al., in press).

TREATMENT ORGANIZATIONS. A variety of treatment methods are available to problem gamblers through organizations and private counselors. For example, Gamblers Anonymous is open to all people who want to stop gambling. At its meetings, which are held throughout the United States, gamblers remain anonymous by using only their first name. The group method offers compulsive gamblers moral support and an accepting environment where they can talk about their experiences and the problems that gambling creates in their life. Gambling is not treated as a vice but as a progressive illness.

In "Changing Spousal Roles in and Their Effects on Recovery in Gamblers Anonymous: GamAnon, Social Support, Wives and Husbands" (*Journal of Gambling Studies*, vol. 26, September 2010), Peter Ferentzy, Wayne Skinner, and Paul Antze explain that Gamblers Anonymous is a society based on "mutual aid," through which gambling addicts recover by following the organization's 12-step recovery program. (See Table 2.7.) These steps are similar to those employed by support groups such as Alcoholics Anonymous. Although the steps have a spiritual aspect, Gamblers Anonymous is not affiliated with any religious group or institution, and the organization is funded by donations. The premise of Gamblers Anonymous is that a recovering compulsive gambler cannot gamble at all without succumbing to the gambling compulsion, so it advocates a "cold turkey" approach to quitting (the gambler just stops gambling) rather than a gradual reduction in gambling activity.

TABLE 2.7

Gamblers Anonymous 12-step recovery program

1. We admitted we were powerless over gambling—that our lives had become unmanageable.
2. Came to believe that a power greater than ourselves could restore us to a normal way of thinking and living.
3. Made a decision to turn our will and our lives over to the care of this power of our own understanding.
4. Made a searching and fearless moral and financial inventory of ourselves.
5. Admitted to ourselves and to another human being the exact nature of our wrongs.
6. Were entirely ready to have these defects of character removed.
7. Humbly asked God (of our understanding) to remove our shortcomings.
8. Made a list of all persons we had harmed and became willing to make amends to them all.
9. Make direct amends to such people wherever possible, except when to do so would injure them or others.
10. Continued to take personal inventory and when we were wrong, promptly admitted it.
11. Sought through prayer and meditation to improve our conscious contact with God as we understood Him, praying only for knowledge of His will for us and the power to carry that out.
12. Having made an effort to practice these principles in all our affairs, we tried to carry this message to other compulsive gamblers.

SOURCE: "Recovery Program," Gamblers Anonymous, 2017, http://www.gamblersanonymous.org/ga/content/recovery-program (accessed June 26, 2017)

The National Council on Problem Gambling is a non-profit organization that was founded to increase public awareness about pathological gambling and to encourage the development of educational, research, and treatment programs. It sponsors the *Journal of Gambling Studies*, an academic journal that is dedicated to scientific research, and operates a confidential hotline (1-800-522-4700) for problem gamblers who need help. The council indicates in "NCPG Affiliate Member List" (March 24, 2014, http://www.ncpgambling.org/files/public/Affiliate_List.pdf) that it has a head office in Washington, D.C., and 36 state affiliate chapters.

The council also directs the National Certified Gambling Counselor program and offers a database of counselors throughout the United States who have completed its certification program. Other organizations that certify gambling counselors include the American Compulsive Gambling Certification Board and the American Academy of Health Care Providers in the Addictive Disorders.

TREATMENT METHODS. Many problem gamblers seek professional counseling. The most common treatment method, in both group and individual counseling sessions, is cognitive behavior therapy. The cognitive portion of the therapy focuses attention on the person's thoughts, beliefs, and assumptions about gambling. According to Timothy W. Fong of the University of California, Los Angeles, in "Pathological Gambling: Update on Assessment and Treatment" (*Psychiatric Times*, vol. 26, no. 9, August 27, 2009), the primary goal is recognizing and changing faulty thinking patterns, such as a belief that gambling can lead to great riches. Behavior therapy focuses on changing harmful behaviors. Most counselors favor complete abstinence from gambling during treatment. For those with mild to moderate gambling problems, treatment usually involves weekly meetings with a support group and/or individual counseling sessions.

Nicki Dowling, David Smith, and Trang Thomas find in "A Comparison of Individual and Group Cognitive-Behavioural Treatment for Female Pathological Gambling" (*Behaviour Research and Therapy*, vol. 45, no. 9, September 2007) that individual treatment seems to be more effective than group treatment at least among some gamblers, although Tony Toneatto and Rosa Dragonetti indicate in "Effectiveness of Community-Based Treatment for Problem Gambling: A Quasi-experimental Evaluation of Cognitive-Behavioral vs. Twelve-Step Therapy" (*American Journal on Addictions*, vol. 17, no. 4, July–August 2008) that 12-step programs are also effective. Those with severe gambling problems usually check into addiction treatment centers to curb their addiction. Such treatment centers isolate patients from the outside world, so they can focus on overcoming their addiction. Some treatment centers even forbid patients from keeping cash on them or from using laptops, phones, or any device that could allow them to gamble.

Bojana Knezevic and David M. Ledgerwood report in "Gambling Severity, Impulsivity, and Psychopathology: Comparison of Treatment- and Community-Recruited Pathological Gamblers" (*American Journal on Addictions*, vol. 21, no. 6, November–December 2012) that pathological gamblers generally exhibit a strong tendency toward impulsive behavior, suggesting that such restrictions are crucial in the treatment process. In "Resilience and Self-Control Impairment" (Sam Goldstein and Robert B. Brooks, eds., *Handbook of Resilience in Children*, 2013), Wai Chen and Eric Taylor study potential links between lack of self-control and attention deficit hyperactivity disorder in gambling and other forms of impulsive activity. Meredith Brown et al. report in "An Empirical Study of Personality Disorders among Treatment-Seeking Problem Gamblers" (*Journal of Gambling Studies*, vol. 32, no. 4, December 2016) that individuals diagnosed with borderline personality disorder (a psychological condition characterized by emotional instability, erratic or inconsistent behavior, impulsivity, and low self-esteem) show greater susceptibility to gambling addiction than individuals who do not suffer from the disease.

Increasingly, mental health professionals and gambling treatment centers are using antidepressants along with cognitive behavior therapy to treat compulsive gambling. Researchers speculate that some compulsive gamblers experience highly elevated levels of euphoria-causing chemicals, such as dopamine, in the brain when they gamble. A number of antidepressant drugs have been proven to prevent such chemicals from interacting with the brain. For example, Jon E. Grant et al. report in "Multicenter Investigation of the Opioid Antagonist Nalmefene in the Treatment of Pathological Gambling" (*American Journal of Psychiatry*, vol. 163, no. 2, February 2006) that the antidepressant nalmefene significantly lowers the need to gamble among people diagnosed with compulsive gambling.

In "Opiate Antagonists in Treatment of Pathological Gambling" (*Brown University Psychopharmacology Update*, vol. 20, no. 3, March 2009), Jon E. Grant et al. find that subjects with a family history of alcoholism have the strongest response to the opiate antagonists that are used to treat pathological gambling. Donald W. Black, Martha C. Shaw, and Jeff Allen of the University of Iowa find in "Extended Release Carbamazepine in the Treatment of Pathological Gambling: An Open-Label Study" (*Progress in Neuro-psychopharmacology and Biological Psychiatry*, vol. 32, no. 5, July 1, 2008) that carbamazepine, an antiseizure drug that is sometimes used to treat bipolar disorder, appears to be effective in treating pathological gambling. Jochen Mutschler et al. report in "Disulfiram, an Option for the Treatment of Pathological Gambling?" (*Alcohol and Alcoholism*, vol. 45, no. 2, December 16, 2010) that disulfiram, which is typically used to treat alcoholism, might also be an effective treatment for pathological gambling. In "Amantadine in the Treatment of Pathological Gambling: A Case Report" (*Frontiers in Psychiatry*, vol. 3, November 2012), Mauro Pettorruso et al. find that the antiviral drug amantadine has the potential to reduce symptoms in pathological gamblers by between 43% and 64%.

CHAPTER 3

AN INTRODUCTION TO CASINOS

When most people imagine a casino, they think of a Las Vegas megaresort consisting of a massive hotel and entertainment complex that is decked out with neon lights, games, and an international crowd of high rollers. Casinos, however, come in all sizes. Besides massive resorts, there are large casinos that serve primarily local customers and do not offer lodging. There are small card rooms and electronic gaming parlors. There are floating casinos—some are functioning boats that take customers on cruises and some are stationary structures built on barges. Casino game machines have been introduced at racetracks to create racinos, and in some states machines are allowed in truck stops, bars, grocery stores, and other small businesses.

The federal government classifies all businesses and industries operated within the United States by using the North American Industry Classification System (NAICS), which issues a six-digit code for each type of business and industry. In *2017 NAICS Definition* (2017, https://www.census.gov/eos/www/naics/), the U.S. Census Bureau defines the NAICS code for casinos, 713210, as follows: "This industry comprises establishments primarily engaged in operating gambling facilities that offer table wagering games along with other gambling activities, such as slot machines and sports betting. These establishments often provide food and beverage services. Included in this industry are floating casinos (i.e., gambling cruises, riverboat casinos)." Casino hotels—that is, hotels with a casino on the premises—fall under the code 721120 and are described as follows: "This industry comprises establishments primarily engaged in providing short-term lodging in hotel facilities with a casino on the premises. The casino on premises includes table wagering games and may include other gambling activities, such as slot machines and sports betting. These establishments generally offer a range of services and amenities, such as food and beverage services, entertainment, valet parking, swimming pools, and conference and convention facilities."

Successful casinos take in billions of dollars each year for the companies, corporations, investors, and Native American tribes that own and operate them. State and local governments also reap casino revenues in the form of taxes, fees, and other payments.

THE HISTORICAL AND CURRENT STATUS OF CASINOS

Gambling was illegal for most of the nation's history. This did not keep casino games from being played, sometimes openly and with the complicity of local law enforcement, but it did keep casinos from developing into a legitimate industry. Even after casino gambling was legalized in Nevada in 1931, its growth outside that state was stifled for decades. It took 47 years before a second state, New Jersey, decided to allow casino gambling within its borders.

As Atlantic City, New Jersey, opened casinos in 1978, a shift occurred in the legality of gambling elsewhere in the country, much of it due to the efforts of some Native American tribes. A string of legal victories allowed the tribes to convert the small-time bingo halls they had been operating into full-scale casinos. Other states also wanted to profit from casino gambling. Between 1989 and 1996 nine other states authorized commercial casino gambling: Colorado, Illinois, Indiana, Iowa, Louisiana, Michigan, Mississippi, Missouri, and South Dakota.

The American Gaming Association (AGA) estimates in *2016 State of the States: The AGA Survey of the Casino Industry* (November 2016, https://www.americangaming.org/sites/default/files/2016%20State%20of%20the%20States_FINAL.pdf) that commercial casinos had revenues of $38.5 billion in 2015, up 2.3% over 2014. This was the industry's highest revenue total on record. In "Gross Gaming Revenue Trending" (July 16, 2017, https://www.nigc.gov/images/uploads/reports/2016GrossGamingRevenueTrends.pdf), the National Indian Gaming Commission indicates that tribal casino revenues totaled $31.2 billion in 2016, the highest total on record, up from $29.9 billion in 2015. According to the

AGA, 460 commercial casinos, 486 tribal casinos, and 55 racetrack casinos (or racinos) operated nationwide in 2015. Additionally, 350 card rooms and 15,423 establishments offering electronic gaming devices were in operation.

In 2015 commercial casinos operated in 18 states: Colorado, Illinois, Indiana, Iowa, Kansas, Louisiana, Maine, Maryland, Michigan, Mississippi, Missouri, Nevada, New Jersey, Ohio, Pennsylvania, Rhode Island, South Dakota, and West Virginia. Tribal casinos operated in 28 states: Alabama, Alaska, Arizona, California, Colorado, Connecticut, Florida, Idaho, Iowa, Kansas, Louisiana, Michigan, Minnesota, Mississippi, Montana, Nebraska, Nevada, New Mexico, New York, North Carolina, North Dakota, Oklahoma, Oregon, South Dakota, Texas, Washington, Wisconsin, and Wyoming. Racinos operated in 14 states: Delaware, Florida, Indiana, Iowa, Louisiana, Maine, Maryland, Massachusetts, New Mexico, New York, Ohio, Oklahoma, Pennsylvania, and West Virginia. Card rooms operated in five states: California, Florida, Minnesota, Montana, and Washington. Seven states offered electronic gaming devices in noncasino locations: Illinois, Louisiana, Montana, Nevada, Oregon, South Dakota, and West Virginia.

CASINO ACCEPTABILITY

The AGA periodically surveys Americans' attitudes toward casino gambling. In *2013 State of the States: The AGA Survey of Casino Entertainment* (2013, https://www.americangaming.org/sites/default/files/research_files/aga_sos2013_rev042014.pdf), the AGA indicates that 47% of respondents considered casino gambling acceptable for anyone and that 38% considered it acceptable for others but not for themselves. Only 14% of respondents thought casino gambling was not acceptable for anyone. Since 2004 the percentage of those who believed casino gambling was acceptable for anyone had dropped from 54%, whereas the percentage of those who said it was acceptable for others but not themselves had risen from 27%. In other words, although the percentage of those who thought gambling was acceptable for themselves or for others in total had not changed, a slight shift had occurred in views of the personal acceptability of gambling.

CASINO GAMES

Casinos offer a variety of games, including card games, dice games, domino games, slot machines, and gambling devices (such as the roulette wheel). Some games are banked games, meaning that the house has a stake in the outcome of the game and bets against the players. Banked games include blackjack, craps, keno, roulette, and traditional slot machines. A nonbanked game is one in which the payout and the house's cut depend on the number of players or the amount that is bet, not the outcome of the game. In percentage games the house collects a share of the amount wagered.

For example, in nonbanked poker players bank their own games. Each player puts money into the "pot" and competes against the other players to win the pot. A portion of the pot is taken by the house. In house-banked poker the players compete against the house rather than each other. Another type of house-banked poker game is one in which there is a posted payout schedule for winning hands rather than for a pot.

Slot machines are by far the most popular form of casino gambling. According to the AGA, in *2013 State of the States*, 61% of casino visitors identified slots as their favorite casino game in 2012, more than triple the proportion who favored blackjack (19%), the second most popular game. Roulette was the game of choice for 8% of casino visitors in 2012, and both poker and craps were identified as the first choice of 4% of visitors.

The AGA explains in "Taking the Mystery out of the Machine: A Guide to Understanding Slot Machines" (July 8, 2010, http://www.oapcg.org/pdfs/taking_the_mystery_out_of_the_machine_brochure.pdf) that casinos utilize two main types of slot machines: mechanical reel machines and video slot machines. Mechanical reel machines have actual moving parts that are set in motion by a player pulling a lever, whereas video machines reproduce a digital version of this mechanical movement onscreen. There are many different subcategories of machines within these two main groupings, including progressive machines whose jackpots increase in size as the player deposits increasing amounts of money and machines that offer various forms of bonus awards related to the appearance of certain symbols. Slot machines can be played for a variety of denominations—from a penny up to more than $5. The $0.25 and $0.50 slot machines are the most popular.

Gaming machines are simple to operate and can offer large payouts for small wagers. Introduced in 1899, the first commercial gambling machines were called slot machines because the gambler inserted a coin into a slot to begin play. Each slot machine consisted of a metal box housing three reels, each of which was decorated all around with symbols (usually types of fruit or spades, hearts, diamonds, and clubs). When the player pulled the handle on the machine, the reels spun randomly until stoppers within the machine slowed them. If a matching sequence of symbols appeared when the reels stopped, the player won. Each reel had many symbols, so literally thousands of outcomes were possible. Because of their construction, ease of play, and low odds, slot machines came to be known as "one-armed bandits."

In the 21st century slot machines are manufactured to strict technical specifications and use a computer programming technique called random number generation.

A computer chip in each machine determines the percentage of payout. Electronic gaming machines offer sophisticated graphics and sound reminiscent of video games. Some are designed to mimic the look and feel of reel-type machines. Patrons may have a choice of a modern push button or an old-fashioned handle to activate play.

Odds against Gamblers

Casinos are businesses whose purpose is to make money, not offer gamblers optimal odds of success. Accordingly, casino games are highly engineered to ensure that the house always comes out ahead of players over time. In "The House Advantage: A Guide to Understanding the Odds" (2011, http://coloradogaming.com/wp-content/uploads/2015/11/odds_brochure_2011_final.pdf), the AGA explains that the size of the house's advantage differs from region to region and from casino to casino. House advantages are expressed in terms of a percentage. In a game that features a house advantage of 10%, the casino expects to win $10 for every $100 that a player wagers. Players of such a game should expect to lose 10% on each wager and to lose more money the longer they continue to play. Although a high level of playing skill can, in certain games and circumstances, reduce the size of the house's advantage, and although some players win more or less than the average due to chance or skill, the house's percentage is constant in the long run. Other factors can increase the house's advantage. For example, a faster than average rate of play always favors the casino.

As Table 3.1 shows, the typical house advantage for individual casino games varies from 1.2% for baccarat to 27% for certain forms of keno (a lottery-like game). In games involving skill or strategy, such as blackjack, some forms of baccarat, and video poker, the house's advantage is calculated assuming an average level of player skill. Slot machines leave no role for player skill or strategy, so the house's advantage remains constant.

Many gamblers harbor a variety of erroneous beliefs about their chances of winning. One of the most commonly observed beliefs is the "gambler's fallacy," a psychological term for the belief that the probability of what will happen next, in a game of chance, changes based on what has happened in the past. For example, many players of slot machines believe a machine that has not paid out for a long time is destined to pay out soon. In fact, the probability that a machine will pay out does not change over time. The same logic applies to all casino games. The odds of rolling a given combination of numbers in craps is the same for each roll of the dice. Several unlucky number combinations in a row do nothing to change the probability of rolling another unlucky combination. The probability of drawing a good hand in poker has nothing to do with the hands one has already drawn.

TABLE 3.1

Casino house advantages, by game

Game	House Advantage
Roulette (double-zero)	5.3%
Craps (pass/come)	1.4%
Craps (pass/come with double odds)	0.6%
Blackjack–average player	2.0%
Blackjack–6 decks, basic strategy*	0.5%
Blackjack–single deck, basic strategy*	0.0%
Baccarat (no tie bets)	1.2%
Caribbean Stud*	5.2%
Let it Ride*	3.5%
Three Card poker*	3.4%
Pai Gow poker (ante/play)*	2.5%
Slots	5%–10%
Video poker*	0.5%–3%
Keno (average)	27.0%

*Optimal strategy

SOURCE: Robert Hannum, "House Advantages for Popular Casino Games," in *Casino Mathematics*, University of Nevada, Las Vegas, Center for Gaming Research, June 2016, http://gaming.unlv.edu/casinomath.html (accessed June 26, 2017)

Streaks of apparent good luck in gambling are nothing of the sort. They are simply a series of random outcomes in which the gambler perceives a pattern. Mathematical probability ensures that such streaks, like streaks of apparent bad luck, will even out over time and that the overall results of play will conform to the size of each individual game's built-in house advantage.

HOW DO CASINOS PERSUADE PEOPLE TO GAMBLE?

Casino gambling is different from other forms of gambling, such as lotteries and Internet gambling, because of its social aspect. Players are either directly interacting with others, as in craps or poker, or surrounded by other people as they play the slot machines. Players often shout out encouragement. Alcoholic drinks are easily accessible and delivered directly to gamblers by waitstaff that circulate throughout the casino. Nonalcoholic drinks and snacks are sometimes provided free of charge. The casino atmosphere is designed around noise, light, and excitement.

Also, casinos arrange slot machines in large groups to attract gamblers. Matthew J. Rockloff, Nancy Greer, and Carly Fay study in "The Social Contagion of Gambling: How Venue Size Contributes to Player Losses" (*Journal of Gambling Studies*, September 25, 2010) the effects of the size of the gambling venue on the behaviors of 135 gamblers. They find that the larger the crowd of gamblers, the faster gamblers played. Although gamblers in larger crowds tended to bet slightly less than gamblers playing alone or in small crowds, increased gambler persistence led to greater gambling losses over the long term.

Most commercial casino visitors participate in activities other than gambling. The AGA notes in *2013 State of the States* that in 2012, 69% of casino visitors ate in a "fine dining restaurant," 55% attended a show or concert, 45% went to a bar or nightclub, 42% went shopping, and 35% used recreational facilities such as a spa, pool, fitness facility, or golf course. In fact, 26% of those surveyed said that when they visited a casino they gambled rarely if at all. Among casino visitors who never or rarely gamble at casinos, 65% said that dining and 39% said that going to shows are their favorite activities when they visit.

Casinos use sophisticated marketing and design to attract gamblers to their facilities and to keep them gambling as long and as happily as possible. They invest millions of dollars to determine which colors, sounds, and scents are the most appealing to patrons. The legend that oxygen is pumped into casinos to keep customers alert is not true—it would be an extreme fire hazard. Nevertheless, most casinos do use bright and sometimes gaudy floor and wall coverings that have a stimulating and cheering effect. Red is a popular decorating color because it is thought to make people lose track of time. Also, clocks typically do not appear on casino walls.

In "Casinos Use TV Stars to Draw New Customers" (Associated Press, April 3, 2010), Wayne Parry explains that during the recession casino marketing campaigns used television and movie celebrities to attract customers. For example, reality television chefs cooked for customers, television shows were filmed at casinos, and reality dance competition contestants danced in shows. According to Liz Benston, in "To Slot Players, Palms' Sideshow a Freak Show" (LasVegasSun.com, September 18, 2010), other casinos began offering more free shows in an attempt to attract customers. With continuing pressures of Internet gambling, casinos have to find more ways to attract customers. Chuck Darrow indicates in "New Live-Entertainment Venues Energize Casinos and Their Patrons" (Philly.com, May 31, 2012) that casinos continue to open entertainment venues and large retail spaces to attract gamblers and other tourists.

Mark D. Griffiths of Nottingham Trent University studies the characteristics of casinos to determine the role that psychology plays in casino design. In "The Money Maze" (PsychologyToday.com, July 18, 2013), he notes that many seemingly simple design choices, such as the location of certain slot machines and the floor layout, can enhance a casino's profitability. For example, restaurants, the most popular nongambling destination for casino visitors, are frequently located so that customers must pass gaming areas on the way to and from their meals. Casinos also frequently arrange floor space in such a way that customers must travel circuitous routes between one area and another, which has the effect of keeping them in the casino longer. Furthermore, casinos manipulate customers' senses to promote gambling. They use lighting at the red end of the electromagnetic spectrum, which has been shown to intensify gambling behavior more than colors at the blue end of the spectrum; they ensure that the casino floor is constantly abuzz with sound to promote feelings of fun and excitement; and some even apply pleasing scents to gaming machines, in accordance with research showing that this boosts slot machine profits. Casinos also take the comfort of their customers seriously by providing ample seating, refreshments, and restrooms so as to encourage players to spend the maximum amount of time gambling.

Casinos also provide numerous perks that reward gamblers who wager the most money. Most casinos offer "comps," or complimentary items such as free or discounted drinks, food, lodging, and entertainment. Casinos concentrate their comp investments on the "high rollers" (gamblers who spend much more than average). Such people often gamble in special rooms, separate from the main casino floor, where the stakes (i.e., the amount bet) can be in the tens of thousands of dollars. Casinos make much of their profit from these high-stakes gamblers. Therefore, the high rollers receive comps that are worth a great deal of money, such as free luxury suites, and are lavished with lots of personal attention.

Less-expensive comps are available to smaller spenders. Most casinos offer clubs that are similar to airline frequent-flyer programs. Gamblers who join receive a card that can be swiped electronically before they play a game. Casino computers track their usage and spending habits and tally up points that can be exchanged for coupons for free slot play or for free or discounted meals, drinks, and shows. The comp programs also serve as a valuable marketing tool for the casinos because they develop a patron database that can be used for advertising and to track trends in game preference and spending.

CHAPTER 4
COMMERCIAL CASINOS

THE MARKET

Commercial casinos are profit-making businesses owned by individuals, private companies, or large public corporations. The term *commercial casino* is used in the United States to indicate a gaming facility that is not owned and operated on Native American lands by a tribal government. (See Chapter 5.) State governments closely regulate casinos. Some states allow land-based casinos, whereas others restrict casino games to floating gambling halls on barges or riverboats. A handful of states allow casino games such as slot machines at other locations, including horse and dog racetracks or other commercial establishments. Some states allow only limited-stakes gambling, in which a limit is placed on the amount that can be wagered.

In *2016 State of the States: The AGA Survey of the Casino Industry* (November 2016, https://www.americangaming.org/sites/default/files/2016%20State%20of%20the%20States_FINAL.pdf), the American Gaming Association (AGA) indicates that in 2015 there were 460 commercial casinos operating in 18 states: Colorado, Illinois, Indiana, Iowa, Kansas, Louisiana, Maine, Maryland, Michigan, Mississippi, Missouri, Nevada, New Jersey, Ohio, Pennsylvania, Rhode Island, South Dakota, and West Virginia. These included both land-based and floating casinos. Major markets for floating casinos included Illinois, Indiana, Iowa, Mississippi, and Louisiana. In addition, 14 states oversaw the operation of 55 racetrack casinos (or racinos) in 2015.

NEVADA

Gambling has a long history in Nevada. It was common in the frontier towns of the Old West but was outlawed at the end of the 19th century. In the early 20th century illegal gambling was widely tolerated throughout the state, and in 1931 gambling was legalized again. The country was in a deep economic depression at the time, and gambling was seen as a source of badly needed revenue.

Casino development was slow at first. Many businesspeople were not convinced that the desert towns of Nevada could attract sufficient tourists to make the operations profitable. In 1941 El Rancho Vegas opened in Las Vegas. Five years later the mobster Bugsy Siegel (1906–1947) opened the Flamingo Hotel and Casino, also in Las Vegas. (Siegel was eventually murdered, reportedly by his business partners because of cost overruns.) Organized crime's relationship with Las Vegas continued for 30 years and tainted casino gambling in many people's minds.

Although the state of Nevada began collecting gaming taxes during the 1940s, regulation of the casinos was lax until the 1970s. Organized crime figures were pushed out of the casino business after Congress passed the Racketeer Influenced and Corrupt Organizations Act in 1970. Corporations moved in to take their place. In 1975 gaming revenues in the state reached $1 billion, according to the Las Vegas Convention and Visitors Authority (LVCVA), in "Stats & Facts: History of Las Vegas" (2017, http://www.lvcva.com/stats-and-facts/history-of-las-vegas).

Many different forms of legal gambling are available in Nevada, including live bingo, keno, and horse racing; card rooms; casino games; and offtrack and phone betting on sports events and horse races. Establishments such as bars, restaurants, and stores are restricted to less than 15 slot machines. Casinos are allowed to have more than 15 machines; many have hundreds.

According to the AGA, in *2016 State of the States*, in 2015 there were 271 commercial casinos operating in Nevada—by far the most of any state. The state's commercial casinos generated $11.1 billion in revenues, employed 170,618 people, and contributed $889.1 million in tax

FIGURE 4.1

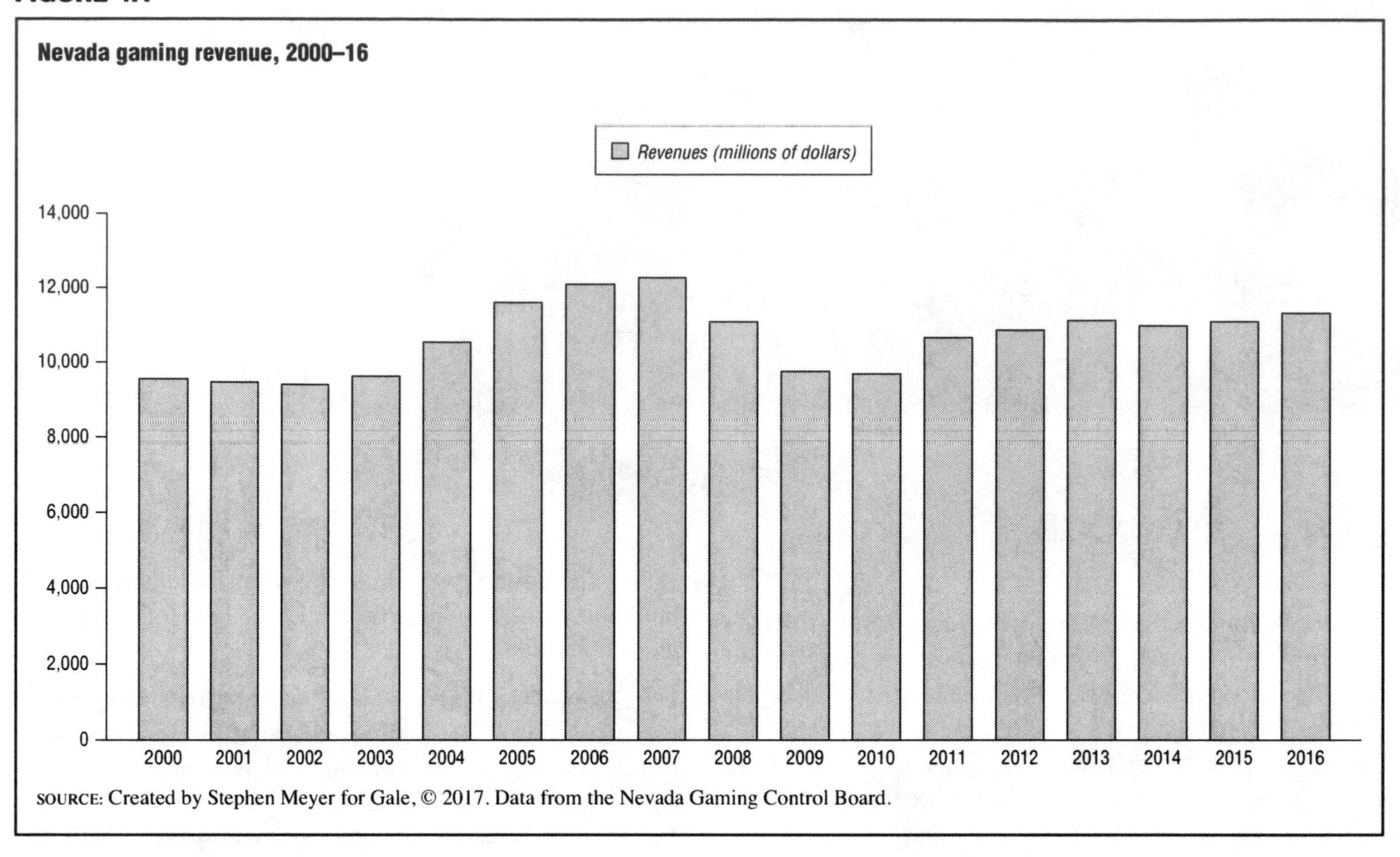

Nevada gaming revenue, 2000–16

SOURCE: Created by Stephen Meyer for Gale, © 2017. Data from the Nevada Gaming Control Board.

revenues to the state's general fund. The Nevada State Gaming Control Board, which releases monthly data on casino revenues in the state, notes that revenues climbed to $11.3 billion in 2016. (See Figure 4.1.) As in other states and industries, the Nevada gaming industry was still rebounding from the effects of the Great Recession, which lasted from late 2007 to mid-2009. Although revenues had increased steadily since 2010, they remained below the record-high of $12.3 billion in 2007.

Nevada casino revenue, or casino win, for 2016 is broken down by gambling category in Table 4.1. Slot machines accounted for $7.2 billion of the casinos' total gaming revenues of $11.3 billion. Table games and other games, including twenty-one (also known as blackjack), craps, roulette, various forms of poker, and sports betting, brought in $4.1 billion in revenues. One-cent slot machines ($3.1 billion) made the most money of any single gaming offering in Nevada's casinos in 2016, followed by multi-denomination slot machines ($2.9 billion), baccarat ($1.2 billion), and twenty-one ($1.1 billion).

Although casinos are located throughout the state, the major gambling markets in Nevada are in the southern part of the state in Clark County (which encompasses Las Vegas and Laughlin) and along the California border in Washoe County (which encompasses Reno) and the Lake Tahoe resort area.

Las Vegas

No other U.S. city is more associated with casinos than Las Vegas. Until 2006, the city was the undisputed gambling capital of the world. However, Kate O'Keeffe reports in "Macau's 2013 Gambling Revenue Rose 19% to $45.2 Billion" (WSJ.com, January 2, 2014) that in 2006 the Chinese territory of Macau surpassed Las Vegas in the size of its gambling revenue, and by 2013 Macau's gambling revenues were $45.2 billion, or five times those of Clark County ($8.7 billion). Although no longer the world's casino capital, and in spite of the rapid growth in casino gambling in other U.S. states since the 1990s, Las Vegas remains the heart of the U.S. commercial casino industry.

The casinos on the 4-mile (6.4-km) stretch of Las Vegas Boulevard known as the Strip make up the top commercial casino market in the country. More than 30 hotel casinos line the Strip, a number of which are among the largest hotels in the United States. These lavishly decorated megaresorts offer amenities such as spas, pools, top-quality restaurants, and top-notch entertainment. The companies operating these establishments generate a substantial amount of their revenues from nongambling sources, including lodging, dining, and entertainment. According to the Nevada State Gaming Control Board, in *Gaming Revenue Report: December 31, 2016* (2017, http://gaming.nv.gov/modules/showdocument.aspx?documentid=11852), Las Vegas Strip casinos generated $6.4 billion in revenues in

TABLE 4.1

Nevada gaming revenue, by type, 2016

[Win amounts are in thousands (add 000)]

Unit description	# of reporting locations—334				
	# of locations	# of units	Win amount	% change	Win percent
Games and tables					
Twenty-one	149	2,617	1,144,719	4.27	13.40
Craps	125	368	384,069	0.33	14.80
Roulette	120	449	351,254	8.91	17.63
3-card poker	105	227	144,497	−3.13	31.60
Baccarat	28	339	1,222,203	−5.28	14.33
Mini-baccarat	47	123	90,387	−10.20	11.61
Keno	54	74	26,828	−2.68	27.68
Bingo	42	42	6,835	−1.70	3.13
Let it Ride	55	75	33,158	−10.92	24.64
Pai Gow	17	18	11,070	−29.40	15.53
Pai Gow poker	92	246	103,256	−0.40	20.60
Race book	137	137	45,703	−6.30	14.73
Sports pool	192	193	219,174	−5.44	4.86
Card games	73	661	117,753	−0.18	
Other games & tables		312	193,085	8.70	24.89
Total games & tables	**164**	**5,881**	**4,093,991**	**−0.40**	**13.82**
Slot machines					
1 cent	239	55,145	3,064,623	7.15	9.91
5 cent	164	2,198	62,925	−25.18	5.90
25 cent	198	6,546	307,027	−8.67	6.61
1 dollar	194	7,594	526,714	2.91	5.90
Megabucks	138	442	52,029	−7.58	11.61
5 dollar	103	1,085	98,394	−0.97	5.54
25 dollar	52	201	29,385	5.20	4.99
100 dollar	33	124	25,798	−0.83	6.28
Multi denomination	319	71,628	2,922,159	−0.14	5.07
Other slot machines		850	74,103	−2.78	
Total slot machines	**327**	**145,813**	**7,163,156**	**2.27**	**6.65**
Total gaming win			**11,257,147**	**1.28**	

Notes: Columns may not foot due to rounding. Unit detail is shown separately only when there are 3 or more locations reporting specific unit information. Otherwise, such information is included in 'Other' category. Figures are current as of: 01/24/17.

SOURCE: Adapted from "Statewide: All Nonrestricted Locations," in *Gaming Revenue Report: December 31, 2016*, State of Nevada, State Gaming Control Board, 2017, http://gaming.nv.gov/modules/showdocument.aspx?documentid=11852 (accessed June 26, 2017)

2016. (See Table 4.2.) By comparison, casinos in the city's downtown area accounted for $564.6 million in revenues that same year. (See Table 4.3.) Other areas of Clark County with revenues placing them among the largest casino markets in the United States include the Boulder Strip area ($788 million), Laughlin ($473.3 million), North Las Vegas ($275.3 million), and Mesquite ($117.8 million). In total, Clark County had 171 casinos in 2016, and their $9.7 billion in total revenues represented nearly 86% of total Nevada gaming revenues.

To a greater degree than other casino markets in the United States, Las Vegas in general and the Strip in particular derive large amounts of revenue from sources other than gambling. Restaurants, theatrical shows, luxury hotels, nightclubs, spas, and golf courses all contribute to the overall economic impact of Las Vegas casinos. At the city's hotel casinos, especially the largest and most luxurious among them, gambling is often one among many vacation activities that visitors access. For example, MGM Resorts International, whose domestic holdings are focused on some of the Strip's most high-end properties, reported $4.7 billion in noncasino revenues (from its hotel rooms, food and beverage sales, and entertainment and retail sales) at its wholly owned domestic resorts in 2016, compared with $3.1 billion in gambling revenues, according to its *2016 Annual Report* (2017, http://mgmresorts.investorroom.com/download/2016+ MGM+Annual+Report.pdf). By comparison, the Boyd Gaming Corporation, whose Las Vegas holdings are geared more toward local and regional visitors and whose other casino holdings are largely in the Midwest and South, derives a much higher proportion of its revenues from gaming. According to Boyd's 10-K filing (2017, http://boydgaming.investorroom.com/download/Boyd+Gaming+2016+10-K.PDF) with the U.S. Securities and Exchange Commission, the company generated $1.8 billion of its revenues from gaming, $306.1 million from food and beverages, and $170.8 million from its hotel rooms in 2016.

The LVCVA releases an annual profile of visitors to the city. In *Las Vegas Visitor Profile Study: 2016* (2017,

TABLE 4.2

Las Vegas Strip gaming revenue, by type, 2016

[Win amounts are in thousands (add 000)]

	# of reporting locations—40				
Unit description	**# of loc's**	**# of units**	**Win amount**	**% change**	**Win percent**
Games and tables					
Twenty-one	37	1,328	875,307	4.50	12.97
Craps	36	183	268,715	0.49	14.59
Roulette	37	262	290,560	10.16	17.04
3-card poker	32	117	102,581	−3.49	32.92
Baccarat	21	318	1,212,452	−5.75	14.42
Mini-baccarat	25	77	69,561	−12.80	12.04
Keno	8	13	4,422	−2.83	27.74
Bingo	3	3	996	−36.80	4.57
Let it Ride	23	37	21,129	−12.57	24.54
Pai Gow	11	11	8,676	−34.98	15.30
Pai Gow poker	31	97	62,152	−0.90	21.17
Race book	36	36	18,649	−9.24	15.67
Sports pool	37	37	96,374	−8.91	3.98
Card games	21	320	78,020	0.88	
Other games & tables		181	143,145	6.88	25.15
Total games & tables	**37**	**3,020**	**3,252,738**	**−0.99**	**13.99**
Slot machines					
1 cent	39	15,132	1,242,513	5.57	11.45
5 cent	33	520	27,828	−18.87	8.29
25 cent	37	2,229	153,546	−4.98	9.68
1 dollar	38	3,221	316,191	3.98	7.17
Megabucks	37	164	25,433	−6.17	12.12
5 dollar	32	558	66,359	−1.15	5.76
25 dollar	27	96	23,055	10.24	5.06
100 dollar	23	84	22,569	1.68	6.09
Multi denomination	40	18,348	1,195,248	0.01	6.23
Other slot machines		393	50,775	−4.49	
Total slot machines	**40**	**40,745**	**3,123,518**	**1.99**	**7.93**
Total gaming win			**6,376,256**	**0.45**	

Notes: Columns may not foot due to rounding. Unit detail is shown separately only when there are 3 or more locations reporting specific unit information. Otherwise, such information is included in 'Other' categories. Figures are current as of: 01/24/17.

SOURCE: Adapted from "Clark County Las Vegas Strip Area: All Nonrestricted Locations," in *Gaming Revenue Report: December 31, 2016*, State of Nevada, State Gaming Control Board, 2017, http://gaming.nv.gov/modules/showdocument.aspx?documentid=11852 (accessed June 26, 2017)

http://www.lvcva.com/includes/content/images/media/docs/C2016-Las-Vegas-Visitor-Profile.pdf), the LVCVA finds that in 2016, 52% of Las Vegas visitors said the primary purpose of their trip was vacation or pleasure, and only 4% said their primary purpose was to gamble. Visitors stayed an average of 3.4 nights in the city and spent an average of $116.15 per night on nonpackage lodging. The typical Las Vegas visitor also spent $318.09 on food and drink, $156.91 on shopping, $96.08 on local transportation, $67.55 on shows, and $35.93 on sightseeing.

NEW JERSEY

In June 1976 New Jersey voters legalized casino gambling in Atlantic City, making it the second state to do so. Casinos began operating in the city in 1978, and for more than a decade it was the primary alternative to Las Vegas for casino visitors. Although Atlantic City saw its prominence decline during the 1990s and the first decade of the 21st century as the legalization of gambling elsewhere gave rise to numerous new competitors, it remains a major player in the casino industry, with yearly gambling revenues approximately half the size of those on the Las Vegas Strip.

Atlantic City

Atlantic City was a popular resort destination during the late 1800s and early 1900s. It was easily accessible by rail, and people visited the beautiful beaches and elegant hotels along the boardwalk, which stretches nearly 5 miles (8 km). During the 1960s the city lost most of its tourist trade to beaches farther south, mainly in Florida and the Caribbean, and the city fell into an economic slump. Casinos were seen as a way to revitalize the city and attract tourists again. The first casino, Resorts International, opened in 1978, followed by Caesars Atlantic and Bally's Park Place in 1979. By 1991 casino gambling was permitted 24 hours per day.

The casinos of Atlantic City have not changed the town into a trendy tourist destination as was originally hoped. In fact, Atlantic City has the reputation of being a

TABLE 4.3

Downtown Las Vegas gaming revenue, by type, 2016

[Win amounts are in thousands (add 000)]

	# of reporting locations—20				
Unit description	**# of loc's**	**# of units**	**Win amount**	**% change**	**Win percent**
Games and tables					
Twenty-one	14	211	47,642	0.09	13.17
Craps	12	34	28,119	−4.18	12.23
Roulette	13	35	14,539	3.60	19.67
3-card poker	12	18	10,390	3.85	28.11
Mini-baccarat	4	8	3,445	−21.92	10.53
Keno	6	6	3,457	8.39	25.93
Let it Ride	9	10	4,734	6.45	22.95
Pai Gow poker	10	13	3,998	−3.94	20.37
Race book	10	10	1,259	−11.21	15.68
Sports pool	11	11	28,289	29.17	5.15
Card games	4	40	3,724	−1.63	
Other games & tables		30	14,299	51.31	17.75
Total games & tables	**14**	**426**	**163,894**	**6.19**	**11.46**
Slot machines					
1 cent	17	4,566	201,285	10.42	11.09
5 cent	14	213	4,917	−16.51	6.87
25 cent	14	990	40,131	−5.50	5.50
1 dollar	13	774	41,265	−1.60	5.06
Megabucks	13	47	5,698	−12.75	11.24
5 dollar	10	68	6,636	−1.89	5.78
Multi denomination	19	3,039	94,996	−0.50	4.83
Other slot machines		105	5,808	−3.94	
Total slot machines	**20**	**9,802**	**400,737**	**3.44**	**7.06**
Total gaming win			**564,631**	**4.23**	

Notes: Columns may not foot due to rounding. Unit detail is shown separately only when there are 3 or more locations reporting specific unit information. Otherwise, such information is included in 'Other' categories. Figures are current as of: 01/24/17.

SOURCE: Adapted from "Clark County Downtown Las Vegas Area: All Nonrestricted Locations," in *Gaming Revenue Report: December 31, 2016*, State of Nevada, State Gaming Control Board, 2017, http://gaming.nv.gov/modules/showdocument.aspx?documentid=11852 (accessed June 26, 2017)

slum with casinos. Industry experts point to two primary factors for this perception: the town has historically relied on day-trippers from the surrounding area (especially New York City and Philadelphia) rather than on long-term vacationers and casino tax revenues have largely funded physical and mental health programs throughout the state rather than being invested in local infrastructure and economic development. All Atlantic City casinos are land based. Although they tend to offer larger hotels and more amenities than those in other states, they are much smaller and less resort-oriented than those of the Las Vegas Strip.

Atlantic City's status as the dominant casino destination for people in the mid-Atlantic region began to be challenged during the first decade of the 21st century as a result of increased competition in neighboring states. Particularly harmful to the Atlantic City market was the legalization of gambling in Pennsylvania. The state licensed its first casinos offering slot machines and other electronic gaming in 2006, and table games were legalized in 2009. By 2013 three casinos were operating in the Philadelphia metro area, which was quickly becoming one of the country's top gambling markets. This dramatic increase in competition coincided with the prolonged period of weakened consumer demand resulting from the Great Recession as well as from the serious property losses that Atlantic City suffered as a result of Hurricane Sandy in October 2012.

As Table 4.4 shows, Atlantic City's 10 casinos had a total win of $2.4 billion in 2016. One of these casinos, the Taj Mahal, discontinued operations in October of that year, leaving nine remaining casinos in Atlantic City as of August 2017. One casino—Borgata—accounted for $833.8 million, or 31.2% of all of Atlantic City casino revenues, in 2016. (See Table 4.5.) Resorts derived the largest proportion of revenues from casino operations, with more than three-quarters (77.7%) of the facility's total revenues coming from gaming. As in the casino industry generally, Atlantic City made the majority of its gambling revenue on slot machines. More than $1.7 billion of the total casino win in 2016 came from slots, and table and other games generated $693.7 million. (See Table 4.6.)

Table 4.7 shows a steady decline in total casino industry jobs in Atlantic City. In 2014 the city's casino industry employed 24,843 people. By 2016 employment had fallen 11.4%, to 22,005.

TABLE 4.4

New Jersey casino gaming revenue, selected statistics, 2015–16

	Casino win		(+/−)	Internet gaming win		(+/−)	Total gaming win		(+/−)
	2016	2015	%	2016	2015	%	2016	2015	%
Bally's AC	$210,710,409	$210,557,585	0.1%	—	—	—	$210,710,409	$210,557,585	0.1%
Borgata	722,772,249	696,217,151	3.8%	47,015,036	45,667,287	3.0%	769,787,285	741,884,438	3.8%
Caesars AC	302,004,633	310,313,801	−2.7%	—	—	—	302,004,633	310,313,801	−2.7%
CIENJ	—	—	—	38,699,334	32,638,153	18.6%	38,699,334	32,638,153	18.6%
Golden Nugget	209,684,168	200,261,054	4.7%	42,249,830	30,968,182	36.4%	251,933,998	231,229,236	9.0%
Harrah's AC	358,068,502	374,315,063	−4.3%	—	—	—	358,068,502	374,315,063	−4.3%
Resorts	173,128,820	162,233,016	6.7%	—	—	—	173,128,820	162,233,016	6.7%
Resorts Digital	—	—	—	31,761,839	6,783,831	(b)	31,761,839	6,783,831	(b)
Tropicana	304,149,289	280,070,455	8.6%	36,983,288	32,822,727	12.7%	341,132,577	312,893,182	9.0%
Current operators	2,280,518,070	2,233,968,125	2.1%	196,709,327	148,880,180	32.1%	2,477,227,397	2,382,848,305	4.0%
Discontinued operator (Taj)	125,494,031	180,269,251	(c)	—	—	—	125,494,031	180,269,251	(c)
Industry total	**$2,406,012,101**	**$2,414,237,376**	**−0.3%**	**$196,709,327**	**$148,880,180**	**32.1%**	**$2,602,721,428**	**$2,563,117,556**	**1.5%**

CIENJ = Caesars Interactive Entertainment NJ

Notes: Internet gaming started in November 2013 and is taxed at 15%. Resorts began Internet gaming operations on February 21, 2015 and Resorts Digital was granted a casino license on August 12, 2015. Trump Taj Mahal closed to the public on October 10, 2016.

SOURCE: "Atlantic City Casino Industry (Includes Closed Casino) Total Gaming Win for the Years Ended December 31, 2016 and 2015," New Jersey Casino Control Commission, 2017, http://www.nj.gov/casinos/about/reports/ (accessed July 18, 2017)

TABLE 4.5

Atlantic City casino industry revenues, by source of revenue, 2016

[Dollars in thousands]

	Bally's AC		Borgata[e]		Caesars AC[a]		Golden Nugget[e]		Harrah's AC		Resorts[a]	
	Revenue	As a % of total revenue	Revenue	As a % of total revenue	Revenue	As a % of total revenue	Revenue	As a % of total revenue	Revenue	As a % of total revenue	Revenue	As a % of total revenue
Casino	$208,803	68.9%	$756,499	70.6%	$299,582	72.4%	$228,276	77.4%	$354,743	64.1%	$171,150	77.7%
Rooms	38,673	12.8%	119,782	11.2%	38,858	9.4%	16,855	5.7%	84,192	15.2%	23,187	10.5%
Food and beverage[b]	45,409	15.0%	146,631	13.7%	58,446	14.1%	31,884	10.8%	88,599	16.0%	16,620	7.6%
Other[c]	10,079	3.3%	48,510	4.5%	16,733	4.1%	18,046	6.1%	26,204	4.7%	9,252	4.2%
Total revenue	**$302,964**	**100%**	**$1,071,422**	**100%**	**$413,619**	**100%**	**$295,061**	**100%**	**$553,738**	**100%**	**$220,209**	**100%**
Less:												
Promotional allowances[d]	77,997	25.7%	237,637	22.2%	112,041	27.1%	76,107	25.8%	134,973	24.4%	55,750	25.3%
Net revenue	$224,967		$833,785		$301,578		$218,954		$418,765		$164,459	

	Tropicana[e]		Discontinued operator (Taj Mahal closed 10/10/16)		Caesars Interactive Entertainment NJ		Resorts Digital Gaming		Industry totals (Includes closed casino)	
	Revenue	As a % of total revenue	Revenue	As a % of total revenue	Revenue	As a % of total revenue	Revenue	As a % of total revenue	Revenue	As a % of total revenue
Casino	$330,764	75.9%	$124,193	72.9%	$25,011	93.9%	$19,900	97.5%	$2,518,921	71.8%
Rooms	55,104	12.7%	28,767	16.9%	0	0.0%	0	0.0%	405,418	11.5%
Food and beverage[b]	33,682	7.7%	9,659	5.7%	0	0.0%	0	0.0%	430,930	12.3%
Other[c]	15,950	3.7%	7,617	4.5%	1,612	6.1%	500	2.5%	154,503	4.4%
Total revenue	**$435,500**	**100%**	**$170,236**	**100%**	**$26,623**	**100%**	**$20,400**	**100%**	**$3,509,772**	**100%**
Less:										
Promotional allowances[d]	91,376	21.0%	47,645	28.0%	79	0.3%	5,352	26.2%	838,957	23.9%[f]
Net revenue	$344,124		$122,591		$26,544		$15,048		$2,670,815	

[a]Figures do not include Internet gaming revenue. Internet gaming at Caesars AC is operated by Caesars Interactive Entertainment (CIENJ) and at Resorts is operated by Resorts Digital Gaming, which hold casino licenses issued by the Casino Control Commission. As such CIENJ's and Resorts Digital Gaming's revenue is reported separately.

[b]Food and beverage revenue for casino-owned outlets only.

[c]Other revenue reflects concert and show tickets, casino-owned spa revenue, casino-owned retail sales, rental income, and other miscellaneous sources.

[d]Promotional allowances are complimentaries given away free in the normal course of the licensee's business.

[e]Figures include Internet gaming revenue.

[f]Does not include promotional expenses. Promotional expenses are complimentaries not offered for sale to patrons in the normal course of a licensee's business. In 2016, promotional allowances and expenses as a percent of total revenue was 27.8%.

SOURCE: "Atlantic City Casino Industry (Includes Closed Casino) Income Statement Revenue by Percentage for the Year Ended December 31, 2016," New Jersey Casino Control Commission, 2017, http://www.nj.gov/casinos/about/reports/ (accessed July 18, 2017)

TABLE 4.6

New Jersey casino gaming revenue, by type, 2015–16

	2016	2015	Growth (decline) $	%
Table & other games				
Blackjack	$252,591,008	$240,702,765	11,888,243	4.9
Craps	81,650,921	80,950,939	699,982	0.9
Roulette	79,111,504	79,295,603	(184,099)	(0.2)
Spanish 21	18,795,967	18,954,834	(158,867)	(0.8)
Three Card Poker	49,520,870	56,715,838	(7,194,968)	(12.7)
Baccarat	2,258,290	4,449,271	(2,190,981)	(49.2)
Mini Baccarat	89,838,305	83,391,621	6,446,684	7.7
Big Six	3,049,933	3,158,219	(108,286)	(3.4)
Keno	247,871	233,168	14,703	6.3
Let It Ride Poker	13,838,267	15,147,985	(1,309,718)	(8.6)
Pai Gow	7,032,530	8,214,479	(1,181,949)	(14.4)
Pai Gow Poker	12,081,413	13,139,423	(1,058,010)	(8.1)
Four Card Poker	14,848,873	14,833,365	15,508	0.1
Sic Bo	159,297	324,922	(165,625)	(51.0)
Caribbean Stud Poker	2,400,831	2,366,429	34,402	1.5
Casino War	1,460,315	1,196,858	263,457	22.0
Double Attack Blackjack	1,712,579	1,123,528	589,051	52.4
Texas Hold'em Bonus Poker	4,064,974	5,184,668	(1,119,694)	(21.6)
Flop Poker	2,349,520	2,857,667	(508,147)	(17.8)
Ultimate Texas Hold'em	3,925,545	4,837,583	(912,038)	(18.9)
Asia Poker	2,287,993	2,267,665	20,328	0.9
Mississippi Stud	4,683,141	4,332,618	350,523	8.1
Criss Cross Poker	5,540,629	3,083,426	2,457,203	79.7
High Card Flush	6,362,171	3,634,947	2,727,224	75.0
Heads Up Hold'em	2,693,112	—	—	N/A
Hold'em 3 Bonus	—	4,939	(4,939)	N/A
Double Draw Poker	284,561	—	—	N/A
Pack's Poker	49,928	—	—	N/A
Electronic Table Games	26,906	518,706	(491,800)	(94.8)
Skill Based Games	—	3,840	(3,840)	N/A
Poker	30,815,678	31,855,623	(1,039,945)	(3.3)
Tournament table & other games	51,710	—	—	N/A
Total table & other games	**693,734,642**	**682,780,929**	**10,953,713**	**1.6**
Slot machines				
.01 and .02 slot machines	873,039,123	851,055,477	21,983,646	2.6
.05 slot machines	61,003,924	75,073,527	(14,069,603)	(18.7)
.25 slot machines	133,066,079	141,937,409	(8,871,330)	(6.3)
.50 slot machines	23,608,597	24,575,402	(966,805)	(3.9)
1.00 slot machines	176,118,534	180,340,198	(4,221,664)	(2.3)
5.00 slot machines	58,210,278	58,462,061	(251,783)	(0.4)
25.00 slot machines	17,077,827	15,594,678	1,483,149	9.5
100.00 slot machines	12,734,646	12,715,478	19,168	0.2
Multi-denominational machines	333,825,414	347,364,886	(13,539,472)	(3.9)
Other slot machines	23,593,033	24,337,331	(744,298)	(3.1)
Total slot machines	**1,712,277,459**	**1,731,456,447**	**(19,178,988)**	**(1.1)**
Grand total	**$2,406,012,101**	**$2,414,237,376**	**(8,225,275)**	**(0.3)**

SOURCE: "Atlantic City Casino Industry (Includes Closed Casino) Total Casino Win for the Years Ended December 31, 2016 and 2015," New Jersey Casino Control Commission, 2017, http://www.nj.gov/casinos/about/reports/ (accessed July 18, 2017)

PENNSYLVANIA

Although Pennsylvania was only opened to commercial gambling in 2006, it has quickly become a leader among casino regions. In 2012 it surpassed New Jersey to become the nation's second-leading gambling state as measured by commercial casino revenues, in large part due to the presence of four casinos located in or near the Philadelphia area, which draw customers from many of the same population centers previously served by Atlantic City.

The Pennsylvania Race Horse Development and Gaming Act, also known as Act 71, was passed in 2004. The law made slot machines legal in the state as a way to raise revenues. The hope was that these revenues would provide Pennsylvania residents with property tax relief as well as fund economic development and attract tourists. The act also created the Pennsylvania Gaming Control Board, which would grant up to 14 gaming facility licenses, regulate and oversee casino gambling in the state, and provide a program for compulsive gamblers.

Act 71 indicated that slot machine licenses would be issued to seven racetracks, five nontrack stand-alone facilities, and two resort hotels. The maximum number

TABLE 4.7

Atlantic City casino industry employment figures, by casino, 2014–16

	2016	2015	2014
Bally's AC	2,486	2,660	2,993
Borgata	5,763	5,784	6,042
Caesars AC[a]	2,952	2,888	2,724
Golden Nugget	2,200	2,164	2,081
Harrah's AC	3,426	3,271	3,663
Resorts[a]	2,047	1,817	1,881
Tropicana	2,951	2,839	2,857
Total for current operators	**21,825**	**21,423**	**22,241**
Discontinued Operators[b]	180	2,192	2,602
Industry totals	**22,005[c]**	**23,615**	**24,843**

[a]Caesars figures include Caesars Interactive Entertainment (CIENJ) and Caesars Enterprise Services (CES, formerly HEC) and Resorts figures include Resorts Digital. Totals do not include employees of leased outlets. NJ based Internet Intermediaries employee totals provided annually by the DGE. In 2016, the total was approximately 186.
[b]Atlantic Club closed to the public on January 13, 2014, reducing total industry employees by 1,655. In September 2014, Revel, Showboat, and Trump Plaza closed to the public, reducing total industry employees by 4,882. Trump Taj Mahal closed to the public on October 10, 2016, reducing total industry employees by 2,068.
[c]16,210 (73.7%) are full-time employees, 3,135 (14.2%) are part-time employees, and 2,660 (12.1%) are considered other employees.

SOURCE: "Atlantic City Casinos (Includes Closed Casino) Employment Statistics for the Three Years Ended December 31, 2016." Table courtesy of the New Jersey Casino Control Commission, 2017, http://www.nj.gov/casinos/about/reports/(accessed July 18, 2017)

TABLE 4.8

Pennsylvania casino revenues, fiscal year 2015–16

Total game revenues fiscal year 2015–16	
Average number of table games in June 2016	1,186
Gross revenue	$839,177,086
Taxes	$119,685,210
Non-banking tables	226
Gross revenue	$58,427,945
Banking tables	937
Gross revenue	$769,500,062
Fully automated electronic tables	10
Gross revenue	$6,471,815
Hybrid tables	13
Gross revenue	$4,777,264
Slot revenues fiscal year 2015–16	
Average number of slot machines in June 2016	26,634
Wagers	$30,267,751,817
Payouts	$27,316,821,057
Promotional plays	$638,018,591
Adjustments	$1,500
Gross terminal revenue	$2,388,658,549
*Taxes and fees	$1,324,629,545
Taxable slot wins per day	
Jul-15	$258.58
Aug-15	$249.65
Sep-15	$238.41
Oct-15	$243.36
Nov-15	$238.34
Dec-15	$242.96
Jan-16	$227.29
Feb-16	$261.30
Mar-16	$257.98
Apr-16	$262.00
May-16	$252.95
Jun-16	$237.76
Local share distribution fiscal year 2015–16	
Statewide	
Slots	$141,355,536
Tables	$16,502,242

*There is $48,495,924 in local share minimum amount included.

SOURCE: "Table Game Revenues FY 2015–16," in *Pennsylvania Control Board 2015–2016 Annual Report*, Pennsylvania Gaming Control Board, 2017, http://gamingcontrolboard.pa.gov/files/communications/2015-2016_PGCB_Annual_Report.pdf (accessed June 27, 2017)

of allowable slot machines in the state was set at 61,000. In 2006 the board issued casino licenses to six horse racing facilities: two were located near Philadelphia, one near Scranton, one near Harrisburg, one near Erie, and one near Pittsburgh. Permanent stand-alone casino licenses were issued to Foxwoods Casino in Philadelphia (which later became Harrah's), Mount Airy Casino Resort in Stroudsburg, Rivers Casino in Pittsburgh, Sands Casino Resort in Bethlehem, and SugarHouse Casino in Philadelphia.

Several more casinos opened in the following three years, including an additional casino in the Philadelphia area and another in nearby Valley Forge. Act 71 was amended in 2010 to allow for table games, which brought the casinos' gaming offerings in line with competitors in nearby Atlantic City. By 2015 Philadelphia was the country's seventh-largest commercial casino market, according to the AGA in *2016 State of the States*. As of 2015, there were 12 Pennsylvania casinos, nine of which were clustered on the eastern side of the state and thus positioned to draw on the densely populated coastal cities of the mid-Atlantic.

According to the Pennsylvania Gaming Control Board (http://gamingcontrolboard.pa.gov/), between fiscal year (FY) 2009–10 (the year that table games were introduced) and FY 2015–16, Pennsylvania's casino revenues grew from $2.2 billion to $3.2 billion, surpassing those of New Jersey, which had been the United States' second-most-prominent commercial casino destination since the 1970s. Although Nevada generated more than triple Pennsylvania's level of revenue, Pennsylvania led the country in the amount of tax revenue generated from legalized gambling. With a 55% tax rate on slot revenues, which accounted for the overwhelming majority of casino revenues, and a 16% tax rate on table games, Pennsylvania has significantly boosted its state government's bottom line even as its casino industry has grown rapidly. During FY 2015–16 table games generated $839.2 million in revenues and brought in $119.7 million in taxes. (See Table 4.8.) Meanwhile, slots generated $2.4 billion in revenues and brought in $1.3 billion in taxes and fees. In "Pennsylvania Leads U.S. in Gaming with Tax Windfall" (Bloomberg.com, May 18, 2012), Romy Varghese notes that the influx of funds from the casino industry allowed Pennsylvania to cut property taxes without eroding school funding, even as the state cut overall spending.

INDIANA

In 1993 the state of Indiana legalized gambling on up to 11 riverboats in specific areas of the state: in the northwestern corner along Lake Michigan; at the southern border along the Ohio River; and around Patoka Lake in the southern part of the state. The Patoka Lake site initially received a riverboat license, but the plan for this area was later vetoed by the U.S. Army Corps of Engineers.

The first riverboat began operation in December 1995 in Evansville. By December 1996 six riverboats were operating in the state. In 2002 new legislation permitted dockside operation of the riverboats in counties that would accept it. Permanent mooring allows patrons to access the casinos anytime during operating hours rather than just during cruise boarding times. The measure was intended to make Indiana casinos more competitive with those in Illinois.

The new law also changed the wagering tax structure from a 22.5% flat tax on adjusted gross receipts to a graduated tax rate of 15% to 35%. A portion of the increased tax revenue is distributed to counties that do not have casinos. The admissions tax, which remained at $3 per person, is split among the state, county, and city—each gets $1 per person.

The Indiana Gaming Commission notes in *Annual Report to Governor Mike Pence* (September 2016, http://www.in.gov/igc/files/FY2016-Annual.pdf) that in 2016 there were 13 riverboat casinos, land-based casinos, and racinos operating in Indiana. Five of the casinos were situated along Lake Michigan and the outskirts of Chicago, and the remaining eight were spread across the central and southern parts of the state. Games allowed included baccarat, blackjack, craps, poker, and roulette. According to the AGA, in *2016 State of the States*, Indiana's Chicago-area casinos, together with a number of casinos on the Illinois side of the border, formed the third-largest casino market in the United States. The Indiana Gaming Commission indicates that the state's total casino win during FY 2016 was more than $2.2 billion and that the state and local governments collected $604.6 million in taxes from casinos. (See Table 4.9.)

MISSISSIPPI

In 1989 Mississippi became the first state to permit gambling on cruise ships that were in state waters on their way to or from international waters. Gambling in Mississippi, however, has had a long history. Gambling along the Mississippi River and its connecting waterways was widespread during the early 1800s. The rivers were the equivalent of the modern-day interstate highway system in that they carried cash-laden farmers, merchants, and tourists to bustling towns along the riverbanks. Gambling halls became notorious establishments that attracted professional gamblers, especially cardsharps, who employed various methods of cheating to earn a living at cards.

By the 1830s the cardsharps had worn out their welcome. According to Richard Dunstan, in *Gambling in California* (1997), five cardsharps were lynched in Mississippi in 1835, so the professional gamblers decided to move to the riverboats that cruised up and down the rivers. Gambling was a popular pastime for riverboat passengers during the 1840s and 1850s. The onset of the Civil War (1861–1865) and then the antigambling movement around the dawn of the 20th century dampened, but did not destroy, open gambling in the state.

During and after World War II (1939–1945) the Mississippi coast experienced a resurgence in illegal casino gambling, particularly in Harrison County, where the Keesler Air Force Base is located. During the 1960s

TABLE 4.9

Indiana casino revenues and taxes, fiscal year 2016

Fiscal year 2016	Total win	Wagering tax	Admission tax	Total tax
Ameristar	$229,658,931.00	$62,975,938.16	$6,980,364.00	$69,956,302.16
Belterra	$109,881,426.00	$23,455,070.52	$3,425,880.00	$26,880,950.52
Blue Chip	$161,957,776.00	$39,270,123.11	$7,080,798.00	$46,350,921.11
French Lick*	$84,421,473.00	$15,751,014.60	N/A	$15,751,014.60
Hollywood	$175,387,198.00	$43,932,618.01	$4,634,205.00	$48,566,823.01
Hoosier Park	$204,001,980.00	$48,878,940.86	N/A	$48,878,940.86
Horseshoe Hammond	$418,542,159.00	$129,186,629.14	$11,146,302.00	$140,332,931.14
Horseshoe Southern	$250,828,424.00	$70,382,568.00	$5,714,199.00	$76,096,767.00
Indiana Grand	$259,153,082.00	$60,543,044.86	N/A	$60,543,044.86
Majestic Star	$90,607,298.00	$17,629,608.93	$2,741,883.00	$20,371,491.93
Majestic Star II	$67,054,291.00	$8,777,957.92	$2,741,883.00	$11,519,840.92
Rising Star	$50,824,715.00	$5,240,819.49	$2,294,979.00	$7,535,798.49
Tropicana Evansville	$126,332,918.00	$28,334,013.89	$3,469,083.00	$31,803,096.89
Totals	**$2,228,651,671.00**	**$554,358,347.48**	**$50,229,576.00**	**$604,587,923.48**

Notes: *Pursuant to Indiana Code 4-33-13-1.5, French Lick Resort-Casino paid an additional 2.5 million dollars in wagering tax by virtue of achieving over 75 million dollars in adjusted gross revenue for fiscal year 2016. The 2.5 million dollar payment is reflected in the wagering tax amounts above.

SOURCE: "FY 2016 Tax Overview," in *2016 Annual Report to Governor Mike Pence*, Indiana Gaming Commission, 2017, http://www.in.gov/igc/files/FY2016-Annual.pdf (accessed June 27, 2017)

the Alcohol Beverage Control Board began refusing licenses to public facilities that allowed gambling. A few private clubs and lodges continued to offer card games and slot machines, but they were shut down by the mid-1980s.

In 1987 the *Europa Star* and several other ships from Biloxi ports began taking gamblers on "cruises to nowhere"—cruises to international waters in the Gulf of Mexico, where passengers could gamble legally. Although the cruises were supported by the city of Biloxi, the state initially opposed them until it became apparent that they were reviving tourism in port towns. Mississippi, which was the poorest state at the time of the 1980 census, was perennially in need of economic development.

In 1990 the Mississippi legislature legalized casino gambling, although each county was allowed to decide whether it would permit gambling within its borders. Fourteen counties along the Gulf Coast and the Mississippi River held referenda to allow dockside casinos, and all voted them down. The next year a city-by-city vote was held, and voters in Biloxi, which was nearly bankrupt at the time, approved the referendum. In 1992 nine dockside casinos opened in Biloxi.

Casinos are grouped in three parts of the state: the northern region centered in Tunica; the central region, which includes Greenville, Vicksburg, and Natchez; and the coastal region, which includes Biloxi, Gulfport, and Bay St. Louis. In *2016 State of the States*, the AGA indicates that in 2015 the Mississippi Gulf Coast was the eighth-largest casino market in the United States and Tunica the 18th largest.

Mississippi has not set a limit on the number of casinos that can be built in the state. Instead, it allows competition to dictate the market size. Before Hurricane Katrina hit the state in August 2005 and devastated the coastal areas, casinos were required to be permanently docked in the water along the Mississippi River and the Mississippi Gulf Coast. The gambling halls of the casinos floated on the water, while their associated lodging, dining, and entertainment facilities were on land. After the hurricane partially or completely destroyed all 12 casinos along the Gulf Coast, the legislature passed a law allowing casino operators that had establishments on the coast in Biloxi, Gulfport, and Bay St. Louis to relocate their casinos 800 feet (244 m) inland so they would be safe from any future storm surges. Along the Mississippi River, the gambling halls sit in slips that have been cut into the riverbank. The Mississippi Band of Choctaw Indians operates the only inland casinos, which are located in Neshoba County.

Between June 2016 and May 2017 Mississippi casinos generated $2.1 billion in revenues. (See Table 4.10.) More than half ($1.2 billion) of these revenues came from the state's Gulf Coast casinos. Casinos in the northern part of the state accounted for $620.5 million in revenues, while casinos in the central region of the state contributed $295.7 million. The state allows round-the-clock gambling with no bet limits.

The table games offered at Mississippi casinos included baccarat, blackjack, craps, keno, and roulette. Table 4.11 provides an overview of table game activity in Mississippi in May 2017. In casino parlance, "drop" refers to the total amount wagered, while "hold" refers to the amount the casino keeps once the game is finished. As Table 4.11 shows, gamblers wagered $151.1 million on table games in Mississippi in May 2017. Of this total, the hold for the state's casinos was $26.6 million, or 17.6%.

According to the Mississippi Gaming Commission, in "Tax Revenues from Gaming" (July 19, 2017, http://www.dor.ms.gov/Business/Documents/Tax%20Revenue%20Fron%20Gaming%20Run%20Date%20July%2019%202017.pdf), the casino industry produces a substantial percentage of the state's annual budget. During FY 2017 (July 2016 to June 2017) casinos paid $252.9

TABLE 4.10

Mississippi casino revenues, by region, June 2016–May 2017

Month	Year	Central	Coastal	Northern	Total
June	2016	$23,811,053.32	$96,147,628.18	$55,504,346.76	$175,463,028.26
July	2016	$25,874,482.16	$105,521,119.24	$58,372,740.01	$189,768,341.41
August	2016	$23,388,559.67	$98,038,734.80	$49,547,916.79	$170,975,211.26
September	2016	$23,444,818.50	$98,499,136.66	$49,701,745.79	$171,645,700.95
October	2016	$23,243,441.96	$99,765,881.35	$50,935,434.98	$173,944,758.29
November	2016	$21,626,710.23	$87,940,811.77	$45,938,470.27	$155,505,992.27
December	2016	$25,832,568.19	$92,281,749.61	$52,128,429.05	$170,242,746.85
January	2017	$22,116,914.86	$99,100,170.43	$44,789,348.79	$166,006,434.08
February	2017	$26,304,365.50	$95,120,542.80	$50,503,019.26	$171,927,927.56
March	2017	$31,272,939.13	$109,965,579.31	$59,959,323.97	$201,197,842.41
April	2017	$25,111,006.59	$99,506,561.19	$51,872,404.97	$176,489,972.75
May	2017	$23,694,630.69	$96,060,745.75	$51,204,051.83	$170,959,428.27
Total		**$295,721,490.80**	**$1,177,948,661.09**	**$620,457,232.47**	**$2,094,127,384.36**

SOURCE: Created by Stephen Meyer for Gale, © 2017. Data from the Mississippi Gaming Commission.

TABLE 4.11

Mississippi casino tables hold, by game and region, May 2017

Table type	Units	Drop	Hold	Hold per table	Hold %
			Central		
Black Jack	65	$7,472,006.00	$1,563,259.00	$24,050.14	20.92%
Craps	13	$2,344,240.01	$499,219.01	$38,401.46	21.30%
Roulette	7	$621,487.50	$141,267.50	$20,181.07	22.73%
Three Card Poker	7	$486,605.00	$98,584.50	$14,083.50	20.26%
MS Stud	7	$777,190.00	$255,709.50	$36,529.93	32.90%
Other	9	$268,046.00	$35,643.00	$3,960.33	13.30%
Overall for region	**108**	**$11,969,574.51**	**$2,593,682.51**	**$24,015.58**	**21.67%**
			Coastal		
Black Jack	268	$44,906,629.80	$5,290,084.10	$19,739.12	11.78%
Craps	36	$17,777,984.75	$3,181,672.75	$88,379.80	17.90%
Roulette	29	$5,968,579.50	$1,419,218.75	$48,938.58	23.78%
Mini Baccarat	20	$13,403,484.88	$1,980,966.63	$99,048.33	14.78%
Let It Ride	5	$670,824.50	$187,034.50	$37,406.90	27.88%
Pai Gow	20	$2,056,995.80	$446,283.58	$22,314.18	21.70%
Three Card Poker	27	$3,979,220.77	$1,229,467.77	$45,535.84	30.90%
Mississippi Stud	20	$3,994,193.00	$881,277.50	$44,063.88	22.06%
Other	30	$5,487,914.84	$959,521.19	$31,984.04	17.48%
Overall for region	**455**	**$98,245,827.84**	**$15,575,526.77**	**$34,231.93**	**15.85%**
			Northern		
Black Jack	135	$19,734,098.45	$3,663,047.45	$27,133.68	18.56%
Craps	29	$12,849,174.25	$2,413,053.75	$83,208.75	18.78%
Roulette	20	$2,217,986.02	$591,186.52	$29,559.33	26.65%
Let It Ride	3	$208,442.00	$58,731.00	$19,577.00	28.18%
Three Card Poker	17	$1,495,534.50	$505,733.50	$29,749.03	33.82%
Mississippi Stud	17	$2,592,997.50	$827,058.41	$48,650.49	31.90%
Other	16	$1,791,222.00	$407,557.00	$25,472.31	22.75%
Overall for region	**237**	**$40,889,454.72**	**$8,466,367.63**	**$35,723.07**	**20.71%**
Overall for state	**800**	**$151,104,857.07**	**$26,635,576.91**	**$33,294.47**	**17.63%**

SOURCE: Adapted from *Mississippi Gaming Commission Tables Hold Report, May 2017* Mississippi Gaming Commission, June 2017, http://www.msgamingcommission.com/files/monthly_reports/0517hold.pdf(accessed June 27,2017)

million in taxes. Approximately $132 million (52%) was paid into the state's general fund, $84.9 million (34%) went to local governments, and the remainder went to retire debt.

LOUISIANA

Like Mississippi, Louisiana has a long gambling history. In 1823, 11 years after Louisiana became a state, its legislature legalized several forms of gambling and licensed six "temples of chance" in the city of New Orleans. Each was to pay $5,000 per year to fund the Charity Hospital and the College of Orleans. The casinos attracted many patrons, including professional gamblers, swindlers, and thieves. In 1835 the legislature repealed the licensing act and passed laws that made gambling hall owners subject to prison terms or large fines.

However, casino-type gambling continued and even prospered throughout the southern part of the state. By 1840 New Orleans had an estimated 500 gambling halls that employed more than 4,000 people but did not pay revenue to the city. Hundreds of professional gamblers who plied the Mississippi River between New Orleans and St. Louis frequented riverboat casinos. When the Civil War broke out, the riverboats were pressed into military service. In 1869 the legislature legalized casino gambling once more, requiring each casino to again pay the state a tax of $5,000.

In *Bad Bet on the Bayou: The Rise of Gambling in Louisiana and the Fall of Governor Edwin Edwards* (2001), Tyler Bridges credits Louisiana gamblers for popularizing craps and poker in the United States during the 19th century. Both were games of chance that had originated in Europe. The Louisiana state lottery began in 1868 but was outlawed in 1895, along with other forms of gambling, after massive fraud was uncovered. Casino gambling went underground and continued to flourish well into the 1960s, thanks to mobsters and political corruption. Two of the state's governors, Earl Kemp Long (1895–1960) and Edwin Edwards (1927–), were well-known gamblers. Edwards reportedly hosted high-stakes gambling games at the governor's mansion; he also went to prison in 2001 for extorting money from people who sought riverboat casino licenses.

In 1991 the state legalized gambling again, authorizing a lottery, casinos, and the operation of video poker machines in restaurants, bars, and truck stops. The legislature authorized operation of up to 15 riverboat casinos in the state; all but those along the Red River were required to make regularly scheduled cruises. The riverboat casinos were required to be at least 150 feet (46 m) long and decorated to look like 19th-century paddleboats. The first riverboat casino, the *Showboat Star*, began operating in 1993.

New Orleans received special permission from the legislature in 1993 to allow a limited number of land-based casinos. In January 1995 Harrah's Entertainment began construction on a casino in the heart of the city. By November 1995 the casino had declared bankruptcy. Following years of negotiations with the state and the city, it reopened in 1999 but threatened bankruptcy again in 2001, blaming the state's $100 million minimum tax. The legislature reduced the tax to $50 million for 2001 and to $60 million for subsequent years to help keep the casino in business.

In April 2001 the legislature ended the so-called phantom cruises of the riverboat casinos, ruling that it would actually be illegal for them to leave the docks. All riverboats were allowed to begin dockside gambling. However, their tax rate was increased from 18.5% to 21.5%.

The state has four major casino markets: Shreveport–Bossier City, New Orleans, Lake Charles, and Baton Rouge. According to the AGA, in *2016 State of the States*, in 2015 the Lake Charles market was the 11th largest in the United States, the Shreveport–Bossier City market was the 14th largest, and the New Orleans market was the 17th largest. A wide variety of games is allowed in Louisiana casinos, including baccarat, bingo, blackjack, craps, keno, poker, roulette, and slot machines. Louisiana riverboat casinos admitted 20.6 million visitors and generated $1.8 billion in gross revenues between July 2016 and May 2017. (See Table 4.12.) As Table 4.13 shows, Louisiana was the nation's third-largest riverboat casino market during FY 2016, after Indiana and Missouri. The Louisiana Gaming Control Board indicates in *21st Annual Report to the Louisiana State Legislature* (2017, http://lgcb.dps.louisiana.gov/docs/2017_Annual_Report.pdf) that the state's land-based casinos generated $293.8 million during FY 2015–16. (See Table 4.14.)

Louisiana is also one of a handful of states that allows video gaming machines to operate in noncasino locations, such as bars, restaurants, hotels, racetracks, and truck stops. As Table 4.15 shows, the state's noncasino video gaming machine revenues were $580.2 million during FY 2015–16.

ILLINOIS

Illinois legalized riverboat gambling in 1990, the second state to do so. The Illinois Gaming Board was authorized to grant up to 10 casino licenses, each of which would allow up to two vessels to be operated at a single specific dock site. Each dock site could not have more than 1,200 gaming positions, and all wagering was to be cashless. Originally, the riverboats were required to cruise during gambling, but they were later allowed to operate dockside. The AGA reports in *2016 State of the States* that two markets in which Illinois casinos had a part—the Chicagoland, Illinois–Indiana market and the St. Louis, Missouri–Illinois market—were among the country's largest, at numbers three and nine, respectively. However, the Illinois casinos shared both markets with the casinos of neighboring states.

TABLE 4.12

Louisiana riverboat gaming statistics, July 2016–May 2017

Riverboat licensees	Opening date	Fiscal year to date admissions	Fiscal year to date total AGR	Fiscal year to date fee remittance
Boomtown Bossier	10/04/96	792,538	54,826,772	11,787,756
Eldorado Resort	12/20/00	1,966,983	120,280,520	25,860,312
Horseshoe	07/09/94	1,246,078	167,597,395	36,033,440
Diamondjacks	05/20/94	687,070	38,499,594	8,277,413
Sam's Town	05/20/04	1,021,553	70,201,576	15,093,339
Margaritaville	06/13/13	1,834,881	134,557,465	28,929,855
Isle Lake Charles	07/29/95	1,138,146	111,794,691	24,035,859
L'auberge Lake Charles	05/23/05	3,194,717	308,300,491	66,284,606
Golden Nugget Lake Charles	12/06/14	3,242,644	240,856,759	51,784,203
Amelia Belle	05/16/07	503,497	38,702,533	8,321,044
Boomtown New Orleans	08/06/94	1,134,009	109,859,381	23,619,767
Treasure Chest	09/05/94	992,416	98,626,287	21,204,652
Belle of Baton Rouge	09/30/94	748,308	57,359,395	12,332,270
Hollywood Baton Rouge	12/28/94	688,777	64,551,637	13,878,602
L'auberge Baton Rouge	09/01/12	1,392,575	160,713,437	34,553,389
Riverboat total		**20,584,192**	**$1,776,727,931**	**$381,996,505**

AGR = adjusted gaming revenue

SOURCE: "Louisiana State Police Fiscal Year-to-Date Activity Summary—Riverboats for the Period of: July 1, 2016–May 31, 2017," in *Riverboat Revenue Reports*, Louisiana Gaming Control Board, June 2017, http://lgcb.dps.louisiana.gov/revenue_reports_riverboat.html (accessed June 27, 2017)

TABLE 4.13

Riverboat casino financial data, by state, fiscal year 2016

	Illinois Calendar year 2016	Indiana Fiscal year 2016	Iowa Fiscal year 2016	Missouri Fiscal year 2016	Mississippi Fiscal year 2016	Louisiana Fiscal year 2016
Number of vessels	10	11	18	13	29	15
Casino type	Dockside/landbased	Dockside/landbased	Dockside/landbased[a]	Dockside	Dockside/landbased[b]	Dockside
AGR	$1,414,431,738	$2,228,651,671	$1,437,136,230	$1,713,132,716	$2,108,317,756	$1,927,763,619
State gaming taxes	$394,926,351	$554,358,347	$301,050,807	$359,757,870	$169,846,926	$414,469,178
Local taxes or fees	$83,066,287	$50,229,576	$14,371,324	$78,485,702	$85,677,731	$83,564,752
Total employees	7,474	13,427	0*	9,371	17,379	14,841
Gaming square footage	359,306	880,743	697,439	849,600	1,315,752	417,966
Total admissions	**12,344,698**	**57,691,761**	**21,538,433**	**42,509,915**	**22,953,331***	**23,284,777**
Tax rate and/or admission charges:	Graduated % of AGR 15%: $0 to $25 M 22.5%: $25 to $50 M 27.5%: $50 to $75 M 32.5% : 75 to $100 M 37.5% : $100 to $150 M 45% : $150 to $200 M 50% : over $200 M Admission tax rates $2 or $3 per admission (based on previous calendar year admission totals.)	Graduated % of AGR 15% $0–$25 M 20% $25–$50 M 25% $50–$75 M 30% $75–$150 M 35% over $150 M $3.00 per admission $4.00 per admission The two newest locations, Hoosier Park (5/29/08) and Indiana Live (6/6/08), do not collect admissions.	State gaming tax rate: 5%: $0–$1 M 10%: $1M–$3 M 22%: over $3 M Number of employees as of the end of fiscal year 2014 is no longer available.	21% of AGR (90%/10% Split—State/home dock community respectively) and & $2.00 per admission ($1.00 to State and $1.00 to Local government.)	8% of AGR to State & 3.2% of AGR to Local governments Mississippi does not track, tax, or charge admission fees. Patron data is gathered through a quarterly survey.	21.5% of AGR

AGR = adjusted gaming revenue

[a]In fiscal year 2015–2016, Iowa began reporting figures as combined totals for its 18 casinos, 1 horse track, and 1 dog track.

[b]Mississippi allows landbased casinos on the Gulf Coast located within 800 feet of a designated waterway.

Note: Information provided by individual states.

SOURCE: "Riverboat Select Data Comparison—Other States," in *21st Annual Report to the Louisiana State Legislature*, Louisiana Gaming Control Board, 2017, http://lgcb.dps.louisiana.gov/docs/2017_Annual_Report.pdf (accessed June 27, 2017)

TABLE 4.14

Louisiana land-based casino financial data, by month, fiscal year 2015–16

	Gross gaming revenue	State tax remittance
July	$35,964,794	$5,081,967
August	22,567,927	5,081,967
September	24,520,784	4,918,033
October	26,519,041	5,081,967
November	22,709,050	4,918,033
December	22,056,708	5,081,967
January	22,508,568	5,081,967
February	25,616,413	4,754,098
March	23,958,789	8,882,966
April	24,600,768	4,931,507
May	23,274,096	5,095,890
June	19,460,894	4,931,507
Total	**$293,757,832**	**$63,841,869**

Note: Information provided by Louisiana State Police, Gaming Audit Section.

SOURCE: "Louisiana Landbased Casino Gaming, FY 2015–2016," in *21st Annual Report to the Louisiana State Legislature*, Louisiana Gaming Control Board, 2017, http://lgcb.dps.louisiana.gov/docs/2017_Annual_Report.pdf (accessed June 27, 2017)

Illinois's 10 riverboat casinos generated $1.4 billion in adjusted gross revenue in 2016, down from $1.6 billion in 2012. (See Table 4.16.) Electronic gaming devices (including slots and other games) generated the majority of Illinois's casino revenues, accounting for $1.1 billion (79.1%) of the total $1.4 billion. (See Table 4.17.)

Illinois levies an admissions tax set at $2 per person at Casino Rock Island and $3 per person for all other casinos, as well as a wagering tax, which starts at 15% for casinos with an adjusted gross revenue of less than or equal to $25 million and increases as revenue increases. The state and communities in which the casinos are located share in casino tax revenues. Illinois's casino industry paid $394.9 million in state taxes ($23.7 million in admission tax and $371.3 million in wagering tax) and $83.1 million in local taxes ($12.3 million in admission tax and $70.7 million in wagering tax) in 2016. (See Table 4.18.)

MISSOURI

The process of bringing casinos to Missouri began in 1992, when 64% of the state's voters approved of a referendum making riverboat gambling legal. That vote was followed by a court case, a constitutional amendment (which was defeated by voters), and a wrangling over the definition of "games of skill." In 1994 the general assembly passed a bill that defined games of skill and authorized riverboats to be located in artificial basins. The first two licenses for riverboat casinos were issued later that year. However, because the casinos could not offer games of chance, such as slot machines, competition from riverboats in Illinois kept customers away, and the casinos were not profitable.

TABLE 4.15

Louisiana video gaming statistics, fiscal year 2015–16

Type of location	No. of VGDs	No. of locations	Dollars in	Dollars out	Net device revenue	Franchise fees
Bars & lounges (Type 1)	2,963	1,028	$286,351,078	$190,892,539	$95,458,540	$24,819,288
Restaurants (Type 2)	1,508	547	114,476,827	75,944,468	38,532,358	10,018,445
Hotels (Type 3)	51	8	4,791,691	3,242,423	1,549,268	402,810
Race tracks & OTBs (Type 4)	993	13	139,802,038	97,469,519	42,332,518	7,619,864
Truck stops (Type 5)	7,645	202	1,314,405,076	912,080,817	402,324,259	130,755,488
Totals	**13,160**	**1,798**	**$1,859,826,710**	**$1,279,629,766**	**$580,196,943**	**$173,615,895**

OTB = Off-track betting
VGD = video gaming device
Notes: Numbers are rounded to the nearest whole dollar. Video gaming devices are classified into five (5) different types according to the establishment in which the devices are located.

SOURCE: "Video Poker Franchise Fees by Type, FY 2015–2016," in *21st Annual Report to the Louisiana State Legislature*, Louisiana Gaming Control Board, 2017, http://lgcb.dps.louisiana.gov/docs/2017_Annual_Report.pdf (accessed June 27, 2017)

TABLE 4.16

Illinois gaming revenues, 2012–16

	2016	2015	2014	2013	2012
Licensees	10	10	10	10	10
Gross receipts	$1,413,478,308	$1,438,029,348	$1,463,418,253	$1,551,311,772	$1,638,167,885
Table games	$294,832,652	$277,942,651	$270,860,357	$267,846,398	$261,811,426
Electronic gaming device	$1,118,645,656	$1,160,086,697	$1,192,557,896	$1,283,465,374	$1,376,356,459
Admissions	12,344,698	12,929,868	13,518,053	14,891,745	16,157,869
Gross receipts per admission	$114.50	$111.22	$108.26	$104.17	$101.39
Total tax	**$477,992,638**	**$488,038,623**	**$500,600,310**	**$536,699,138**	**$574,336,421**
Wagering tax	$441,980,380	$450,284,717	$461,118,691	$493,226,411	$527,161,053
Admission tax	$36,012,258	$37,753,906	$39,481,619	$43,472,727	$47,175,368
State share	$394,926,351	$403,167,011	$413,813,942	$444,205,433	$476,246,146
Local share	$83,066,287	$84,871,612	$86,786,368	$92,493,705	$98,090,275

Note: Gross receipts do not reflect chip float, expired vouchers and prior period adjustments.

SOURCE: "Five Year History," in *2016 Annual Report*, Illinois Gaming Board, 2017, https://www.igb.illinois.gov/FilesAnnualReport/2016IGBAnnualReport.pdf (accessed June 27, 2017)

TABLE 4.17

Illinois gaming revenues, by casino and type, 2016

Docksite	Gross receipts	Table games	% of total	Electronic gaming device	% of total
Alton Belle	$49,119,883	$3,489,518	7.1%	$45,630,365	92.9%
E. Peoria Par-A-Dice	82,442,601	14,026,127	17.0%	68,416,474	83.0%
Casino Rock Island	75,609,430	6,033,960	8.0%	69,575,470	92.0%
Joliet-Hollywood	121,237,915	15,218,330	12.6%	106,019,585	87.4%
Metropolis-Harrah's	80,333,667	15,403,017	19.2%	64,930,650	80.8%
Joliet-Harrah's	183,657,280	33,920,251	18.5%	149,737,029	81.5%
Aurora-Hollywood	120,364,816	19,226,786	16.0%	101,138,030	84.0%
E. St. Louis Casino Queen	109,305,723	17,195,828	15.7%	92,109,895	84.3%
Elgin Grand Victoria	163,513,828	34,616,809	21.2%	128,897,019	78.8%
Des Plaines Rivers Casino	427,893,165	135,702,026	31.7%	292,191,139	68.3%
Total	**$1,413,478,308**	**$294,832,652**	**20.9%**	**$1,118,645,656**	**79.1%**

SOURCE: "Sources of Revenue: Table Games and Electronic Gaming Devices," in *2016 Annual Report*, Illinois Gaming Board, 2017, https://www.igb.illinois.gov/Files AnnualReport/2016IGBAnnualReport.pdf (accessed June 27, 2017)

After a petition drive, voters passed an initiative that allowed "only upon the Mississippi River and the Missouri River, lotteries, gift enterprises, and games of chance to be conducted on excursion gambling boats and floating facilities." The result was significant: revenues from casino riverboats during the first quarter of FY 1996 were more than twice what they had been during the first quarter of the previous year. Initially, the casinos were only allowed to hold two-hour gambling excursions. In 2000 the law was changed to allow continuous boarding. However, the original $500 loss limit per excursion that had been approved in 1992 still applies. Patrons are allowed to purchase only $500 worth of chips or tokens in any two-hour period, preventing them from losing more than that amount within the "excursion" period.

State law limits the number of Missouri riverboat casinos to 13. As of 2017, four of these casinos operated in the St. Louis area and four operated in the Kansas City area. The remaining five casinos were docked in the cities of Boonville, Cape Girardeau, Caruthersville, La Grange, and St. Joseph. Games allowed include blackjack, poker, and other card games, craps, roulette, slot machines, and several wheel games. Combined, the 13

TABLE 4.18

State vs. local share of Illinois gaming taxes, 2015–16

	State share of gaming taxes			Local share of gaming taxes		
Distribution	Admission tax	Wagering tax	Total	Admission tax	Wagering tax	Total
2016	$23,667,560	$371,258,791	$394,926,351	$12,344,698	$70,721,589	$83,066,287
2015	$24,824,038	$378,342,973	$403,167,011	$12,929,868	$71,941,744	$84,871,612
% change	−4.66%	−1.87%	−2.04%	−4.53%	−1.70%	−2.13%

SOURCE: Adapted from "Distribution of Gaming Taxes," in *2016 Annual Report*, Illinois Gaming Board, 2017, https://www.igb.illinois.gov/FilesAnnualReport/2016IGBAnnualReport.pdf (accessed June 28, 2017)

TABLE 4.19

Missouri casino gaming financial data, by casino location, fiscal year 2016

Casino/Location	Location opening date	Admissions	Admission fees[a]	Adjusted gross receipts	Gaming tax[b]	Estimated capital investment[c]	Employees	Table games	Slot machines
Ameristar Casino St. Charles	27-May-94	5,987,697	$11,975,394	$262,230,777	$55,068,463	$819,503,874	1,579	77	2,326
Argosy Riverside Casino	22-Jun-94	3,406,467	$6,812,934	$151,176,202	$31,747,003	$76,272,930	784	36	1,506
St. Jo Frontier Casino	24-Jun-94	1,154,918	$2,309,836	$39,645,908	$8,325,641	$43,136,190	237	10	509
Harrah's North Kansas City	22-Sep-94	3,678,081	$7,356,162	$172,856,596	$36,299,885	$442,303,390	953	61	1,330
Lady Luck—Caruthersville	27-Apr-95	890,889	$1,781,778	$37,140,674	$7,799,542	$59,425,700	228	9	557
Isle of Capri—Kansas City	18-Oct-96	2,341,485	$4,682,970	$77,434,563	$16,261,258	$99,148,518	361	17	977
Ameristar Casino Kansas City	16-Jan-97	5,208,895	$10,417,790	$199,917,227	$41,982,618	$460,670,702	1,053	72	2,150
Hollywood Maryland Heights	11-Mar-97	5,427,689	$10,855,378	$228,481,716	$47,981,160	$83,177,217	1,131	58	2,127
Isle of Capri—Boonville	06-Dec-01	1,815,462	$3,630,924	$81,560,863	$17,127,781	$120,474,000	419	20	914
Mark Twain Casino	25-Jul-01	949,546	$1,899,092	$36,080,407	$7,576,886	$28,738,039	221	12	613
Lumiere Place	19-Dec-07	3,853,669	$7,707,338	$138,139,798	$29,009,358	$272,435,857	936	73	1,762
River City Casino	01-Mar-10	5,930,474	$11,860,948	$223,661,473	$46,968,909	$475,230,386	1,044	55	1,947
Isle of Capri—Cape Girardeau	30-Oct-12	1,864,643	$3,729,286	$64,806,511	$13,609,367	$132,599,314	425	26	930
Grand totals		**42,509,915**	**$85,019,830**	**$1,713,132,716**	**$359,757,870**	**$3,113,116,117**	**9,371**	**526**	**17,648**

[a]Fifty (50) % of admission fees go to the state, with the remaining 50% to the local home dock.
[b]Ninety (90) % of gaming taxes go to the state, with the remaining 10% to the local home dock.
[c]Figures reflect the current property investments since company purchase as reported to MGC (Missouri Gaming Commission).
Note: Figures are subject to adjustment.

SOURCE: Adapted from "Fiscal Year 2016 Project Summary," in *Annual Report 2016*, Missouri Gaming Commission, 2017, http://www.mgc.dps.mo.gov/annual_reports/AR_Current Year/00_FullReport.pdf (accessed June 28, 2017)

Missouri riverboat casinos generated $1.7 billion in revenues during FY 2016. (See Table 4.19.) The Ameristar Casino in St. Charles reported the highest revenue total of all the Missouri casinos that year, with $262.2 million in adjusted gross receipts, followed by the Hollywood Casino in Maryland Heights ($228.5 million) and the River City Casino in St. Louis ($223.7 million). According to the AGA, in *2016 State of the States*, the St. Louis casino market (which Missouri shares with Illinois) was the nation's ninth largest in 2015, with $1 billion in revenues. The Kansas City casino market (which the state shares with Kansas) was 13th that year, with $782.1 million in revenues.

MICHIGAN

Pari-mutuel horse racing (betting in which those who bet on the top competitors share the total amount bet and the house gets a percentage) was legalized in Michigan in 1933. During the 1970s the state lottery was legalized, and a concerted effort was made to allow casino gambling in Detroit. The casino effort was unsuccessful until 1994, when the Windsor Casino opened just across the river in Windsor, Ontario. By that point, more than a dozen tribal casinos were operating around the state of Michigan, and Detroit was in an economic downturn. Attitudes toward casino gambling changed, and in November 1996 Michigan voters narrowly approved ballot Proposal E, which authorized the operation of up to three casinos in any city that had a population of 800,000 people or more and was located within 100 miles (161 km) of any other state or country in which gaming was permitted. Casino gaming also had to be approved by a majority of voters in the city. Proposal E was subsequently modified and signed into law in 1997. Out of 11 casino proposals submitted, three were accepted: Atwater/Circus Circus Casino (later called MotorCity Casino), owned by Detroit Entertainment; Greektown Casino, owned by the Sault Ste. Marie Tribe of Chippewa Indians; and the MGM Grand, owned by MGM Grand Detroit Casino. The casinos were granted permission to open at temporary locations, with permanent facilities planned for a proposed waterfront casino district.

The first casino, MGM Grand, opened in July 1999 in a former Internal Revenue Service building. Later that year the MotorCity Casino started operations in a former bread factory. The Greektown Casino opened in November 2000 in the heart of the city. It was the first tribally owned casino to open off a reservation. Detroit became the largest city in the country to allow casino gambling.

The plan for a casino district was eventually abandoned because of rising real estate prices and local opposition, and the number of hotel rooms initially proposed was cut back after marketing studies showed that many casino customers were regional and did not need overnight lodging. The permanent casinos were also delayed by several lawsuits. However, work began on the permanent MGM Grand and MotorCity Casino during the summer of 2006. The MGM Grand opened in October 2007. The MotorCity Casino opened in stages throughout 2007, with a grand opening in June. The Greektown Casino filed for bankruptcy protection in May 2008. However, it opened an expanded casino in August 2008 and opened its hotel in February 2009.

The Michigan Gaming Control Board reports that the Detroit casinos grossed $1.4 billion in 2016. (See Table 4.20.) The highest monthly earnings total was in March of that year, when the city's casinos generated nearly $125 million in revenues; by comparison, Detroit casinos generated only $108.6 million in revenues in June, the lowest monthly total for the year. Each casino paid 8.1% of adjusted gross receipts as a state wagering tax to be deposited in Michigan's School Aid Fund. Up until 2007 MGM Grand, MotorCity, and Greektown paid an additional 4%; 3.5% went to the state's general fund and 0.5% went to the Agriculture Equine Industry Development Fund. By 2008 MGM Grand and MotorCity were no longer required to pay the additional 4% because both had opened permanent casinos and hotels. As Table 4.21 shows, the combined 8.1% state wagering taxes deposited in the School Aid Fund totaled $112.2 million in 2016.

Unlike casinos in some other states, the Detroit casinos are not permitted under the Michigan Liquor Control Code to provide free alcoholic drinks. Games offered at the Detroit casinos include baccarat, blackjack, casino war, craps, keno, poker, roulette, slot machines, and video poker.

In 2013 the city of Detroit filed for bankruptcy, becoming the largest U.S. municipality ever to do so, and the local tax revenues generated by its three casinos were an integral part of the bankruptcy process. Karen Pierog and Joseph Lichterman report in "Detroit's Path into and out of Bankruptcy Is Paved with Casino Money" (BusinessInsider.com, January 20, 2014) that the city had used its casino tax revenues (one of the city's most stable sources of funding) as collateral for debt in 2009, when its financial troubles began pushing it toward bankruptcy. Detroit's obligations to its creditors kept the casino funds locked up from that time, and they remained inaccessible to the city as it attempted to negotiate a deal with creditors and exit bankruptcy in 2014. Meanwhile, the cash-strapped city was unable to deliver satisfactory services to its residents. Pierog and Lichterman quote Kevyn Orr, the emergency manager for Detroit, as saying, "Every day that we don't have access to casino revenue, we cannot make the necessary investment in this city to provide for the health, safety and welfare of the citizens."

IOWA

Gambling was outlawed in the state of Iowa from the time of its statehood in 1846 until 1972, when a provision in the state constitution that prohibited lotteries was repealed. In 1973 the general assembly authorized bingo

TABLE 4.20

Detroit, Michigan, casino revenues, 2016

Month	MGM Grand casino	MotorCity casino	Greektown casino	Total Detroit casinos
January	$46,718,389	$37,132,789	$25,215,520	$109,066,698
February	47,193,904	40,385,060	28,822,029	116,400,993
March	51,818,121	43,700,909	29,430,087	124,949,117
April	50,462,058	42,346,725	29,548,118	122,356,901
May	51,903,470	39,010,274	27,621,012	118,534,756
June	47,555,020	35,895,882	25,161,439	108,612,341
July	50,185,322	40,569,698	27,707,524	118,462,544
August	51,333,426	37,835,364	26,734,163	115,902,953
September	48,479,768	37,034,304	26,653,492	112,167,564
October	47,347,683	37,182,514	26,207,798	110,737,995
November	48,191,386	37,236,627	26,074,020	111,502,033
December	50,937,207	39,526,855	26,443,809	116,907,871
Total	**$592,125,754**	**$467,857,001**	**$325,619,011**	**$1,385,601,766**

SOURCE: "Casino Revenues," in *2016 Michigan Gaming Control Board Annual Report*, Michigan Gaming Control Board, 2017, http://www.michigan.gov/documents/mgcb/2016_MGCB_Annual_Report_Public_Version_FINAL_558221_7.pdf (accessed June 28, 2017)

TABLE 4.21

Detroit, Michigan, casino state taxes, 2016

Month	MGM Grand casino	MotorCity casino	Greektown casino	Total Detroit casinos
January	$3,784,190	$3,007,756	$2,042,457	$8,834,403
February	3,822,706	3,271,190	2,334,584	9,428,480
March	4,197,268	3,539,774	2,383,837	10,120,879
April	4,087,427	3,430,085	2,393,397	9,910,909
May	4,204,181	3,159,832	2,237,302	9,601,315
June	3,851,957	2,907,566	2,038,076	8,797,599
July	4,065,011	3,286,145	2,244,310	9,595,466
August	4,158,007	3,064,664	2,165,467	9,388,138
September	3,926,861	2,999,779	2,158,933	9,085,573
October	3,835,162	3,011,784	2,122,832	8,969,778
November	3,903,502	3,016,167	2,111,996	9,031,665
December	4,125,914	3,201,675	2,141,949	9,469,538
Total	**$47,962,186**	**$37,896,417**	**$26,375,140**	**$112,233,743**

SOURCE: "State Casino Wagering Tax," in *2016 Michigan Gaming Control Board Annual Report*, Michigan Gaming Control Board, 2017, http://www.michigan.gov/documents/mgcb/2016_MGCB_Annual_Report_Public_Version_FINAL_558221_7.pdf (accessed June 28, 2017)

and raffles by specific parties. A decade later pari-mutuel wagering at dog and horse tracks was legalized, followed by a state lottery in 1985. In 1989 gambling aboard excursion boats was authorized for counties in which voters approved gambling referenda. Between 1989 and 1995 referenda authorizing riverboat gambling were approved in more than a dozen counties. The Iowa Racing and Gaming Commission granted licenses for riverboat gambling in 10 counties: Clarke, Clayton, Clinton, Des Moines, Dubuque, Lee, Polk, Pottawattamie, Scott, and Woodbury. By law, the residents of these counties vote every eight years on a county referendum that allows riverboat gambling to continue. In 1994 pari-mutuel racetracks gained approval to operate slot machines.

In 2016 two of Iowa's riverboat casinos—the Rhythm City Casino in Davenport and the Isle of Capri in Bettendorf—moved their operations to land-based facilities. As of 2017, 17 casinos and two racinos operated in Iowa. Games included bingo, blackjack, craps, keno, minibaccarat, poker, roulette, slots, and video poker. The casinos are required by law to meet space requirements for nongamblers and to provide shopping and tourism options. Slot machines are allowed at racinos only if a specific number of live races are held during each racing season. In *2016 Annual Report* (2017, https://irgc.iowa.gov/sites/default/files/documents/2017/03/annual_report_2016.pdf), the Iowa Racing and Gaming Commission indicates that in 2016 combined admissions to the state's casinos and racinos totaled 21.8 million.

In 2016 gaming revenues totaled $1.4 billion at Iowa's casinos and racinos. (See Table 4.22.) The Iowa Racing and Gaming Commission notes that in 2016 the state collected $288.6 million in gaming taxes from all gaming operations, city and county governments collected another $14.5 million in taxes, and $11.6 million was deposited in an endowment fund. Iowa's gaming tax rate ranges from 5% to 24%, depending on revenue and the type of venue.

TABLE 4.22

Iowa casino financial data, 2015 and 2016

	2015	2016
Admissions	21,651,881	21,774,455
Slot drop	$8,868,050,696	$8,924,419,159
Slot coin in	$13,666,844,908	$13,724,712,684
Slot revenue	$1,282,345,322	$1,299,290,428
Table drop	$645,200,269	$678,215,780
Table revenue	$142,006,148	$146,873,287
Adjusted gross revenue	$1,424,351,470	$1,446,163,715
City tax	$7,121,740	$7,230,798
County tax	$7,121,740	$7,230,798
Endowment fund	$11,394,783	$11,569,281
State miscellaneous fund	$2,848,696	$2,892,322
Gaming tax	$284,169,103	$288,578,149
Regulatory fee	$16,692,078	$14,842,515

SOURCE: "All Iowa Gaming Totals," in *2016 Annual Report*, Iowa Racing and Gaming Commission, 2017, https://irgc.iowa.gov/sites/default/files/documents/2017/03/annual_report_2016.pdf (accessed June 28, 2017)

WEST VIRGINIA

Since 1994 West Virginia has allowed "video lottery," or electronic gaming devices including slot machines, at four racetracks in the state, subject to local referenda. Three racetracks introduced electronic gaming that year, and a fourth followed in 1997. In 2007 the West Virginia legislature passed the Racetrack Table Games Act, which opened the door for legalized casino table games at the racetracks. Two of the tracks introduced table games that year, a third began offering table games in 2008, and the fourth followed in 2010. In 2008 West Virginia voters approved a referendum allowing table games at the Greenbriar Resort in White Sulphur Springs. The resort began offering both electronic gaming devices and table games in 2009, making it the state's only casino not affiliated with a racing facility.

As Figure 4.2 shows, electronic gaming revenues at West Virginia's racetracks fell steadily between 2007 and 2011. After experiencing a slight uptick in 2012, gaming revenues once again experienced a steep decline. Although the addition of table games and the opening of the Greenbriar Resort casino during this period led to increases in the size of table game revenues, total gaming revenues of $581.2 million in 2016 remained well below the revenues reported in 2007.

COLORADO

During the 1800s gambling halls and saloons with card games were prevalent throughout the mining towns of Colorado. Gambling, however, was outlawed in the state around the dawn of the 20th century.

In November 1990 Colorado voters approved a constitutional amendment that permitted limited-stakes gaming in the towns of Black Hawk and Central City, near Denver, and Cripple Creek, near Colorado Springs. The first Colorado casinos opened in October 1991 and had gross revenues of nearly $8.4 million during their first month of operation. Before July 2009 only blackjack, poker, and slot machines were permitted in Colorado's casinos, and the maximum single bet was $5. In November 2008, however, Colorado voters approved Amendment 50, which gave voters in the locales where casinos were operating the option to add craps and roulette and to raise the maximum single bet to $100. Voters in the three locales approved the changes, which went into effect in July 2009.

Colorado's casinos had gross revenues of $810.8 million in 2016, the state's highest total since a peak of $816.1 million in 2007. (See Table 4.23.)

OHIO

In November 2009 Ohio voters approved of an amendment to the state's constitution that authorized the opening of

FIGURE 4.2

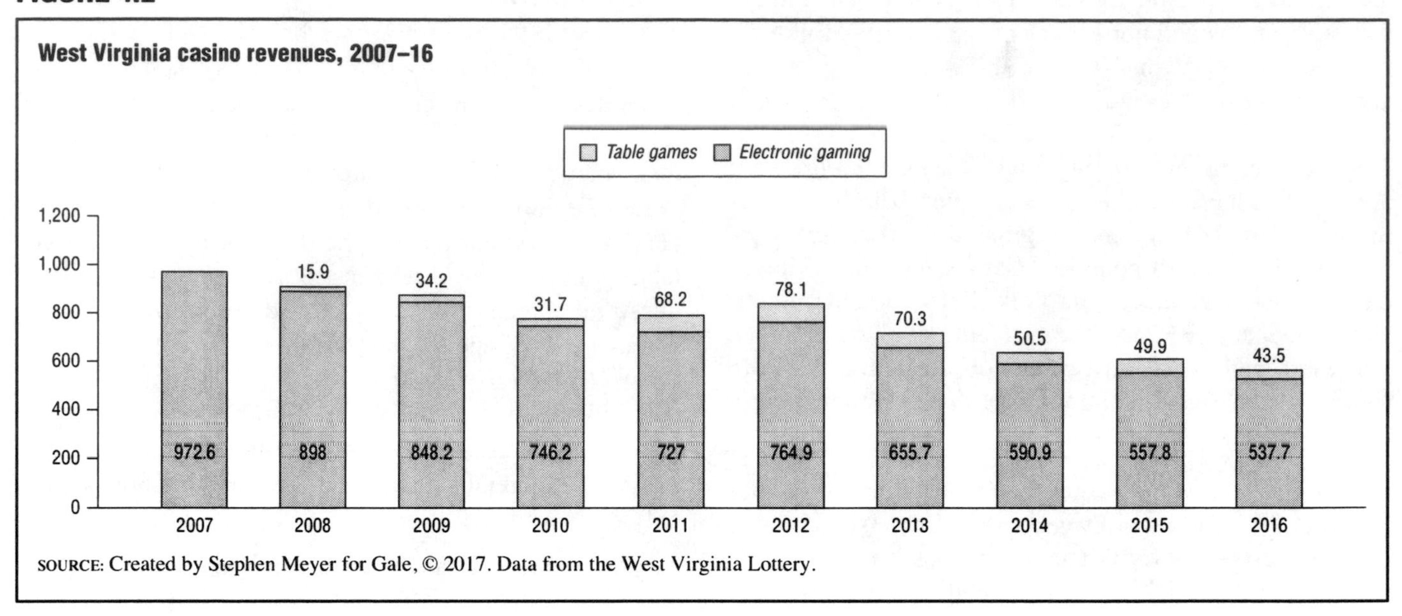

SOURCE: Created by Stephen Meyer for Gale, © 2017. Data from the West Virginia Lottery.

TABLE 4.23

Colorado casino revenues, 1999–2016

	Statewide	% chg	Cripple Creek	% chg	Black Hawk	% chg	Central City	% chg
1999	$551,319,150	15.0%	$122,611,399	8.3%	$354,913,835	30.5%	$73,793,917	−21.5%
2000	$631,852,149	14.6%	$134,630,255	9.8%	$433,768,948	22.2%	$63,452,945	−14.0%
2001	$676,674,192	7.1%	$138,617,688	3.0%	$478,326,427	10.3%	$59,730,077	−5.9%
2002	$719,701,403	6.4%	$142,436,212	2.8%	$524,464,856	9.6%	$52,800,335	−11.6%
2003	$698,910,864	−2.9%	$142,525,101	0.1%	$506,476,837	−3.4%	$49,908,926	−5.5%
2004	$725,903,556	3.9%	$148,689,335	4.3%	$524,035,343	3.5%	$53,178,879	6.6%
2005	$755,499,720	4.1%	$151,011,042	1.6%	$531,878,276	1.5%	$72,610,402	36.5%
2006	$782,098,818	3.5%	$153,075,257	1.4%	$554,484,627	4.3%	$74,538,934	2.7%
2007	$816,129,779	4.4%	$154,962,066	1.2%	$581,385,160	4.9%	$79,782,553	7.0%
2008	$715,879,711	−12.3%	$140,081,962	−9.6%	$508,685,618	−12.5%	$67,112,131	−15.9%
2009	$734,590,354	2.6%	$140,356,304	0.2%	$529,976,828	4.2%	$64,257,223	−4.3%
2010	$759,610,323	3.4%	$134,437,711	−4.2%	$559,445,467	5.6%	$65,727,144	2.3%
2011	$750,108,903	−1.3%	$131,405,587	−2.3%	$550,883,660	−1.5%	$67,819,656	3.2%
2012	$766,254,008	2.2%	$133,160,559	1.3%	$558,542,208	1.4%	$74,551,241	9.9%
2013	$748,707,912	−2.3%	$128,032,315	−3.9%	$553,082,797	−1.0%	$67,592,801	−9.3%
2014	$746,347,749	−0.4%	$123,432,964	−3.6%	$560,598,893	1.4%	$62,315,892	−7.8%
2015	$790,078,238	5.9%	$128,046,346	3.7%	$595,792,993	6.3%	$66,238,898	6.3%
2016	$810,793,527	2.6%	$131,393,766	2.6%	$609,754,552	2.3%	$69,645,209	5.1%
Total	**$15,674,989,730**		**$3,107,104,235**		**$10,787,524,442**		**$1,780,361,052**	

SOURCE: Adapted from "Colorado Casino AGP by Calendar Year," in *Industry Statistics–Gaming*, Division of Gaming, Colorado Department of Revenue Enforcement Division, 2017, https://www.colorado.gov/pacific/sites/default/files/CYStats_1.pdf (accessed June 28, 2017)

casinos in Cincinnati, Cleveland, Columbus, and Toledo. The state's legislature passed the Casino Control Law the following year, laying the groundwork for the implementation of the constitutional amendment and establishing the regulatory framework governing the industry. The state's first casinos, one in Cleveland and one in Toledo, opened in May 2012. Casino gambling arrived in Columbus later that year, and in 2013 a casino opened in Cincinnati. According to the Ohio Casino Control Commission, in "2016 Monthly Casino Revenue Report" (February 7, 2017, http://casinocontrol.ohio.gov/Portals/0/Revenue%20Reports/2016/DECEMBER%202016%20Casino%20Revenue%20AMENDED%20Feb2017.pdf), the state's four casinos generated combined revenues of $797.9 million during the 2016 calendar year—$263.4 million from table games and $534.5 million from slots. In *2016 Annual Report* (September 2016, http://casinocontrol.ohio.gov/Portals/0/Communications/2016%20OCCC%20Annual%20Report%20-%20FINAL%20small.pdf), the Ohio Casino Control Commission notes that the state's casinos pay a 33% tax on gross revenues, and during FY 2016 (July 2015 to June 2016) contributed $270.4 million to state and local governments.

MARYLAND

Maryland legalized electronic gaming (or video lottery terminals, as they are officially called) in 2007. The

Maryland Lottery and Gaming Control Agency reports in *Comprehensive Annual Financial Report for the Years Ended June 30, 2016 and 2015* (2017, http://msa.maryland.gov/megafile/msa/speccol/sc5300/sc5339/000113/021800/021814/20170189e.pdf) that five casinos offering slots and other forms of electronic gaming (including the purchase of actual lottery tickets) had opened by 2016. In 2013 Maryland's casinos began offering table games as a result of the passage of the Gaming Expansion Act by the state assembly. By the end of FY 2016 (June 30) there were 424 table games at the state's casinos. The total FY 2016 gaming revenues in Maryland stood at $741.7 million.

KANSAS

According to Stephanie Simon, in "(State) House Rules in Kansas Casino" (WSJ.com, February 4, 2010), Dodge City, Kansas, was known during the 1880s as "the wickedest little city in America." Among its attractions were casinos, saloons, and brothels. Gambling, however, was outlawed by 1900. More than a century later, in March 2007, it was reinstated when the Kansas legislature passed a bill that allowed the Lottery Commission to establish up to four state-owned casinos as a way to raise revenue and provide jobs for thousands of workers. The law requires the state to own all the gambling equipment, although the buildings are owned by private companies.

The first casino to open was Boot Hill Casino and Resort in Dodge City in December 2009. It was followed by the Kansas Star Casino in Mulvane in 2011 and the Hollywood Casino at the Kansas Speedway in Kansas City in 2012. The Kansas Racing and Gaming Commission indicates in *2016 Annual Report* (2017, http://www.krgc.ks.gov/images/ stories/pdf/Annual_reports/2016_ Annual_Report_Final.pdf) that the three casinos' total revenues were $364.4 million during calendar year 2016.

MAINE

In 2004 Maine's voters approved the legalization of two slot machine casinos in the state, pending local approval. The state's first casino, Hollywood Slots, opened in Bangor in 2005, but local approval did not materialize for a second facility. In 2010 voters approved an expansion of legal gambling in the state, allowing for the opening of a second facility, the Oxford Casino in Oxford, which offered both slots and table games, as well as the addition of table games at Hollywood Slots. These changes went into effect in 2012. According to the Maine Gambling Control Board, in "Revenue Totals" (2017, http://www.state.me.us/dps/GambBoard/revenue-information.html), between January and July 2017 the two casinos generated a total of $710.9 million in gross revenues.

SOUTH DAKOTA

Commercial casino gambling in South Dakota is restricted to the town of Deadwood in Lawrence County. A rustic mountain town about 60 miles (97 km) from Mount Rushmore, Deadwood was designated as a National Historic Landmark and is listed on both the National and South Dakota Registers of Historic Places. The games allowed are blackjack, poker, and slot machines.

The rocky history of gambling in Deadwood is described by Katherine Jensen and Audie Blevins in *The Last Gamble: Betting on the Future in Four Rocky Mountain Mining Towns* (1998). The gold rush of 1876 brought large numbers of people into the town, and it soon became packed with saloons and gambling halls. The town became associated with notorious characters such as Wild Bill Hickok (1837–1876); Alice Ivers Tubbs (1851–1930), commonly called Poker Alice; and Martha Jane Burk (1852?–1903), better known as Calamity Jane.

Although gambling was outlawed in the Dakota Territory in 1881, it continued quite openly in Deadwood with the apparent complicity of the local sheriff. According to Jensen and Blevins, gambling opponents complained in 1907 that the town's gambling halls "operated as openly as grocery stores, running twenty-four hours a day." However, on a busy Saturday night in 1947 the attorney general of South Dakota finally sent 16 raiders into the bars of Deadwood to show the town that the state meant business. The blatant days of gambling were over in Deadwood, although Jensen and Blevins note that the establishments continued to operate quietly for the next four decades.

In 1984 a group of Deadwood businesspeople and community leaders began working to bring legalized gambling back to Deadwood, primarily to raise money to preserve the town's historic buildings. The group developed the slogan "Deadwood You Bet" and had it printed on hundreds of buttons. Despite widespread local support, the idea failed at the ballot box in 1984 and was voted down by the legislature in 1988. The measure made it onto the ballot in November 1988 following a massive petition effort. In 1989 South Dakota voters approved limited-stakes casino gambling for Deadwood. Originally, the casinos could only offer a $5 maximum bet. This limit was raised to $100 in 2000. The South Dakota Commission on Gaming (http://dor.sd.gov/Gaming/Frequently_Asked_Questions) indicates that as of 2017 the betting limit was $1,000.

In *Annual Report—FY 2016* (2016, http://dor.sd.gov/Gaming/Annual_Report/PDFs/FY2016AnnualReport.pdf), the South Dakota Commission on Gaming reports that Deadwood's casinos had record-high gross revenues of $109.2 million during FY 2016 (July 2015 to June 2016), up from $106 million during FY 2015. (See Figure 4.3.) The casino revenues had been growing steadily over the preceding decade, and although growth flattened somewhat

FIGURE 4.3

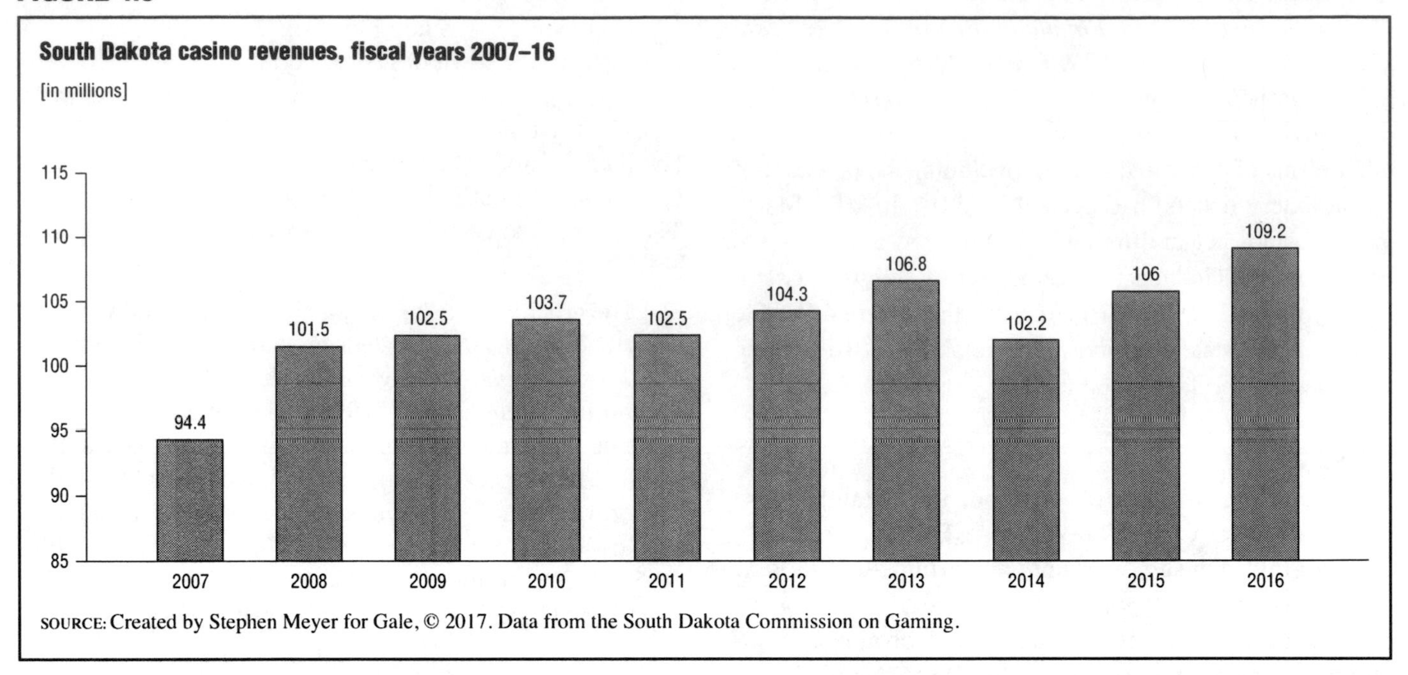

SOURCE: Created by Stephen Meyer for Gale, © 2017. Data from the South Dakota Commission on Gaming.

in the years following the onset of the Great Recession, the overall growth trend continued. The casinos pay an 8% gaming tax on their adjusted gross revenue and a device tax of $2,000 per card game or slot machine. During FY 2016 the tax on gross revenues amounted to $9.3 million and the device tax amounted to $6.4 million.

CHAPTER 5

NATIVE AMERICAN TRIBAL CASINOS

Casinos operated by Native American tribes made $31.2 billion in 2016. (See Figure 5.1.) By comparison, the American Gaming Association (AGA) reports in *2016 State of the States: The AGA Survey of the Casino Industry* (November 2016, https://www.americangaming.org/sites/default/files/2016%20State%20of%20the%20States_FINAL.pdf) that commercial casinos made $38.5 billion in 2015.

According to the U.S. Census Bureau (December 2013, http://www.census.gov/population/www/cen2010/cph-t/t-6tables/TABLE%20(1).pdf), 5.2 million people in the United States (approximately 1.7% of the total U.S. population) identified themselves as Native American or Alaskan Native in the 2010 census. The Census Bureau categorizes native people according to tribal groupings that encompass various specific tribes and bands that share cultural traditions and ethnic origins. In some cases these tribal groupings are specific to a particular geographic region, whereas in other cases the individual tribes and bands are spread across different reservations and U.S. states. As of 2010, the largest of these native subgroups were the Cherokee tribal grouping, with 819,105 members living primarily in the southeastern United States and Oklahoma; the Navajo Nation tribal grouping, with 332,129 members living primarily in Arizona, Utah, and New Mexico; the Choctaw tribal grouping, with 195,764 members located primarily in Oklahoma, Mississippi, Alabama, Louisiana, and Texas; the Mexican American tribal grouping, with 175,494 members living primarily in the southwestern United States; the Chippewa tribal grouping, with 170,742 members located primarily in Minnesota and Wisconsin; the Sioux tribal grouping, with 170,110 members located primarily in North and South Dakota, Nebraska, Minnesota, and Iowa; the Apache tribal grouping, with 111,810 members located primarily in Oklahoma, Texas, Arizona, and New Mexico; and the Blackfeet Tribe of the Blackfeet Indian Reservation of Montana, which numbered 105,304.

As of 2017, the Bureau of Indian Affairs (BIA; http://www.bia.gov), an agency of the U.S. Department of the Interior, officially recognized 567 tribes located within the borders of the United States. Federally recognized tribes are regarded as having a government-to-government relationship with the United States. They enjoy tribal sovereignty, including self-government within certain limits. As such, casinos operated by tribes are subject to different forms of regulation, with the National Indian Gaming Commission (NIGC), an independent federal government agency, taking the lead role in overseeing the industry, whereas states take the lead role in overseeing the commercial casino industry. According to the NIGC, in the press release "2016 Indian Gaming Revenues Increase 4.4 Percent" (July 17, 2017, https://www.nigc.gov/news/detail/2016-indian-gaming-revenues-increased-4.4), 244 tribes located in 28 states operated 484 gaming facilities in 2016.

HISTORY

The growth of tribal casinos can be traced to the late 1970s, when Native American tribes began operating bingo halls to raise funds for tribal purposes. Bingo games were legal in those states but subject to restrictions on the size of the jackpot and how often games could be held. The Seminole Tribe of Florida and the Oneida Tribe of Wisconsin, which had been prevented by state law from holding high-stakes bingo games, took their respective states to court, arguing that they were sovereign nations and as such were not subject to state limitations on gambling.

In 1981 the U.S. Fifth Circuit Court of Appeals ruled in *Seminole Tribe of Florida v. Butterworth* (658 F.2d 310) that the Seminole Tribe could operate a high-stakes bingo parlor because the state of Florida did not have regulatory power over the tribe, which was a sovereign governing entity. A similar ruling was issued in *Oneida Tribe of Indians v. State of Wisconsin* (518 F.Supp. 712 [1981]). Both cases concluded that the states' gambling

FIGURE 5.1

Growth in tribal gaming revenues, 2007–16

[Dollars in billions]

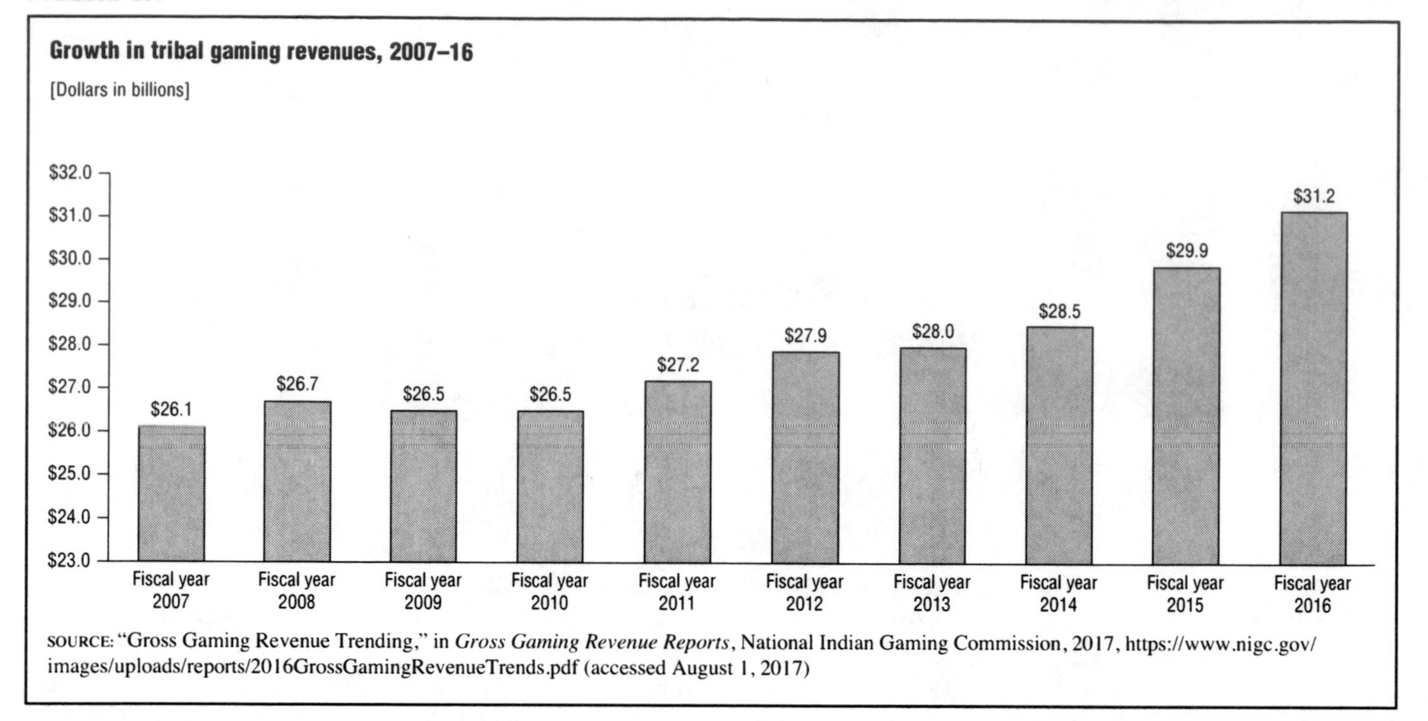

SOURCE: "Gross Gaming Revenue Trending," in *Gross Gaming Revenue Reports*, National Indian Gaming Commission, 2017, https://www.nigc.gov/images/uploads/reports/2016GrossGamingRevenueTrends.pdf (accessed August 1, 2017)

laws were regulatory, or civil, in nature rather than criminal because the states already allowed bingo games to take place.

Other tribes also initiated lawsuits, and the issue eventually reached the U.S. Supreme Court. In *California v. Cabazon Band of Mission Indians* (480 U.S. 202 [1987]), the court ruled that California could not prohibit a tribe from conducting activities (in this case, high-stakes bingo and poker games) that were legal elsewhere in the state. In 1989 the Bay Mills Indian Community opened the King's Club in Brimley, Michigan, the first Native American gambling hall to offer slot machines and blackjack.

GAMBLING CLASSES

In 1988 Congress passed the Indian Gaming Regulatory Act (IGRA) in response to the court decisions. The act allows federally recognized tribes to open gambling establishments on their reservations if the state in which they are located already permits legalized gambling. It set up a regulatory system and three classes of gambling activities:

- Class I—social gaming for minimal prizes and traditional gaming (e.g., in tribal ceremonies or celebrations)
- Class II—bingo and bingolike games, lotto, pull tabs (paper tickets that have tabs concealing symbols or numbers), tip jars (lotterylike games played with preprinted tickets), punch boards (thick cardboard with symbols or numbers concealed behind foil), and nonbanking card games (such as the type of poker that is played against other players instead of the house)
- Class III—banking card games (card games in which the player bets against the house), casino games, slot machines, pari-mutuel betting (in which those who bet on the top competitors share the total amount bet and the house gets a percentage) on horse and greyhound racing and on jai alai, electronic facsimiles of any game of chance, and any other forms of gaming not included in Class I or II

Class I games are regulated exclusively by the tribes and require no financial reporting to other authorities. Class II and III games are allowed only if such games are already permitted in the state where the tribe is located. According to the U.S. Government Accountability Office (GAO), the investigatory branch of Congress, court rulings have maintained that tribes can operate casinos where state-run lotteries exist and charitable casino nights are permitted.

Class II and III operations require that the tribe adopt a gaming ordinance that is approved by the NIGC. In addition, Class III gaming requires that the tribe and state have an agreement, called a tribal-state compact (or treaty), that is approved by the U.S. secretary of the interior. A compact is supposed to balance the interests of the state and the tribe in regard to standards for operation and maintenance, the applicability of state and tribal laws and regulations, and the amount needed by the state to defray its regulatory costs. Tribes may have compacts with more than one state and may have different compacts for different types of gambling operations.

REGULATION

Native American casinos are regulated at three levels of government: federal, state, and tribal. Federal regulation is performed by the NIGC, which oversees the licensing of gaming employees and management and reviews tribal gaming ordinances. The NIGC also has enforcement powers, with penalties ranging from fines to the closure of operations. Most violations do not result in closure, but in notification followed by fines. For example, Jay Weaver reports in "Seminole Tribe Accused of Violating Federal Laws" (MiamiHerald.com, June 8, 2010) that in June 2010 the NIGC charged the Seminole Tribe in Florida with violating gaming laws by spending hundreds of thousands of dollars in gambling profits on six members. The case was resolved in October 2010, with the Seminole Tribe admitting to the violations. In "Agreed to Civil Fine Assessment" (October 27, 2010, https://www.nigc.gov/images/uploads/enforcement-actions/CFA-10-01.pdf), the NIGC indicates that the tribe agreed to pay a fine of $500,000 and to establish stricter auditing procedures.

Most violations concern tribes' or casino operators' failure to follow requirements related to auditing, paperwork, or contractual matters. For example, a September 2013 settlement (https://www.nigc.gov/images/uploads/enforcement-actions/SA121Sep1913.pdf) between the NIGC and individuals affiliated with the Thlopthlocco Tribal Town's Golden Pony Casino in Okemah, Oklahoma, concerned the casino management's failure to complete an approved management contract with the commission. Similar issues were behind the commission's investigation of and eventual settlement with the managers of the Flandreau Santee Sioux Tribe's Royal River Casino in South Dakota (July 20, 2011, https://www.nigc.gov/images/uploads/enforcement-actions/FlandreauSantteSA1101.pdf) and the Ponca Tribe of Oklahoma's Two Rivers Casino (May 13, 2013, https://www.nigc.gov/images/uploads/enforcement-actions/SA-PTO.pdf).

The federal government also has criminal jurisdiction over cases involving federal offenses, such as embezzlement, cheating, and fraud, at tribal gaming operations.

State regulation is spelled out in the tribal-state compacts. They cover matters such as the number of slot machines that may be operated; limits on the types and quantities of card games that can be offered; minimum gambling ages in the casinos; authorization for casino workers to unionize; public health and safety issues; compulsive gambling issues; the effects of tribal gaming on other state enterprises; and how much revenue should be paid to the state and how often.

The tribes themselves are the primary regulators of tribal gaming. Individual tribes that operate casinos typically employ their own tribal gaming commissions to regulate operations in accordance with the IGRA and NIGC standards.

FEDERAL RECOGNITION

Native American casinos must be a tribal endeavor, not an individual endeavor—that is, a random group of Native Americans cannot start a tribal casino. Only a tribe's status as a sovereign entity, which is granted by the federal government, allows it to conduct gaming.

The list of federally recognized tribes is maintained by the BIA. As of August 2017, the most current list was available in "Indian Entities Recognized and Eligible to Receive Services from the United States Bureau of Indian Affairs" (*Federal Register*, vol. 82, no. 10, January 17, 2017).

Throughout U.S. history tribes have received federal recognition through treaties with the U.S. government, via congressional actions, or through BIA decisions. Most tribes were officially recognized during the 18th and 19th centuries. In the 21st century recognition can be achieved either through an act of Congress or through a series of actions, known as the "federal acknowledgment process," that can take many years. Under the Code of Federal Regulations, Title 25, Part 83, Section 83.11 (August 21, 2017, http://www.ecfr.gov/cgi-bin/ECFR?page=browse), a group of Native Americans must meet seven criteria to be federally recognized as a tribe:

- The group must have been identified as a Native American entity on a substantially continuous basis since 1900.
- A predominant portion of the group must make up a distinct community and have existed as a community from historical times to the present.
- The group must have maintained political influence or authority over its members as an autonomous entity from historical times until the present.
- The group must submit a copy of its current governing documents, including membership criteria.
- The group's membership must consist of individuals who descended from a historical Native American tribe or from historical Native American tribes that combined and functioned as a single autonomous political entity.
- The membership of the group must be composed primarily of people who are not members of an existing acknowledged Native American tribe.
- The tribe must not be the subject of congressional legislation that has terminated or forbidden a federal relationship.

Federal recognition is important to Native American tribes if they are to assert their rights to self-government and tribal sovereignty and to be eligible for various forms of federal assistance. According to the BIA, in "What We Do" (August 21, 2017, http://www.bia.gov/WhatWeDo/index.htm), in 2017 the federal government held about

55 million acres (22.3 million ha) of land in trust for federally recognized Native American tribes and their members. If a tribe does not have a land base, the federal government can take land in trust for the tribe once it receives recognition. That land is no longer subject to local jurisdiction, including property taxes and zoning ordinances. The BIA operates an education system that consists of 183 schools and dormitories and 28 tribal colleges, universities, and postsecondary schools. It also offers social services, natural resources management, economic development programs, administration of tribal courts, settlement of land and water claims, housing services, disaster relief, and infrastructure maintenance and replacement.

Most tribes require that a person have a particular degree of Native American heritage (usually 25%) to be an enrolled member. Some tribes require proof of lineage. As of 2017, the BIA estimated the total population of the 567 federally recognized tribes it served at 1.9 million. This differs from the previously cited Census Bureau estimate because the bureau allows individuals to self-identify their status as a Native American or Alaskan Native independently of formal tribal membership statistics.

One of the most contentious issues related to tribal casinos is the authenticity of the tribes themselves. Critics charge that some Native American groups want federal recognition only as a means to enter the lucrative gambling business. The GAO examined this issue in 2001, at which time there were 193 tribes with gambling facilities. The agency reports in *Indian Issues: Improvements Needed in Tribal Recognition Process* (November 2001, http://www.gao.gov/new.items/d0249.pdf) that 170 (88%) of the tribes could trace their federal recognition to the time of the Indian Reorganization Act of 1934 or similar legislation from the 1930s. About 59% of those tribes were engaged in gambling operations in 2001. By contrast, 45% of the tribes recognized since 1960 were engaged in gambling operations.

The GAO indicates that the procedures established by the BIA in 1978 to ensure that the recognition of tribes be uniform and objective had become too long and inconsistent. Backlogs became constant because the number of petitions for recognition began to climb during the 1990s. However, in *Indian Issues: Timeliness of the Tribal Recognition Process Has Improved, but It Will Take Years to Clear the Existing Backlog of Petitions* (February 10, 2005, http://www.gao.gov/new.items/d05347t.pdf), the GAO explains that by 2005 the backlog of cases had been steadily reduced. Other sources, such as the article "Landless Tribe Waits Federal Recognition" (Associated Press, January 5, 2008), reported in 2008 that waits were as long as 15 years. As of 2017, the most recent tribe to have received recognition was the Pamunkey Tribe of King William County, Virginia. Wayne Covil reports in "Pamunkey Indians Become First Virginia Tribe Given Federal Recognition" (WTVR.com, February 2, 2016) that the Pamunkey Tribe lived in the region when the first English settlers arrived in 1607 and were the first Native American tribe from the state of Virginia to receive federal recognition.

In "Bureau of Indian Affairs Proposes Revising Rules for Recognizing Native American Tribes" (WashingtonPost.com, May 25, 2014), Michael Melia notes that the BIA announced in May 2014 proposed changes to the federal recognition process to increase its transparency and efficiency. A key change included replacing the requirement that tribes must be able to show political authority dating from "historical times" with a requirement to show political authority dating to 1934, the year Congress first acknowledged tribes as political entities. Prior to that date, many tribes preserved their histories only through oral means, in keeping with their longstanding cultural traditions, and many had likewise fought numerous wars and been the victims of extermination campaigns, which made the preservation of written records regarding political matters difficult at best. The new rules set specific percentages for the proportion of a group's membership that must be descended from the tribe since historical times (80%) and for the proportion that must make up a community (30%). They also proposed to allow groups previously denied recognition to reapply. The new rules were subject to a period of public comment prior to being finalized.

REVENUES

Because tribes are sovereign governments, they are not required by law to make public statements of their revenues, including casino revenues, so financial information on individual tribal casinos is not publicly released. Each year the NIGC announces total gaming revenue from the previous year for all tribal gaming facilities combined, and it typically provides an accompanying assortment of selected data related to tribal gaming. As shown in Figure 5.1, tribal casinos had revenues of $31.2 billion in 2016, up 19.5% from 2007, when revenues stood at $26.1 billion.

The NIGC operates seven regional offices across the United States, each of which oversees gaming on tribal lands within a multistate area. (See Figure 5.2.) The Portland, Oregon, office oversees tribal gaming in Alaska, Idaho, Oregon, and Washington; the Sacramento, California, office oversees tribal gaming in California and northern Nevada; the Phoenix, Arizona, office oversees tribal gaming in southern Nevada, Arizona, Colorado, and New Mexico; the St. Paul, Minnesota, office oversees tribal gaming in Montana, Wyoming, North and South Dakota, Nebraska, Minnesota, Iowa, Wisconsin, and Michigan; the Tulsa, Oklahoma, office oversees tribal gaming in Kansas and eastern Oklahoma; the Oklahoma City office oversees tribal gaming in western Oklahoma and Texas; and the Washington, D.C., office oversees tribal gaming in Louisiana, Mississippi,

FIGURE 5.2

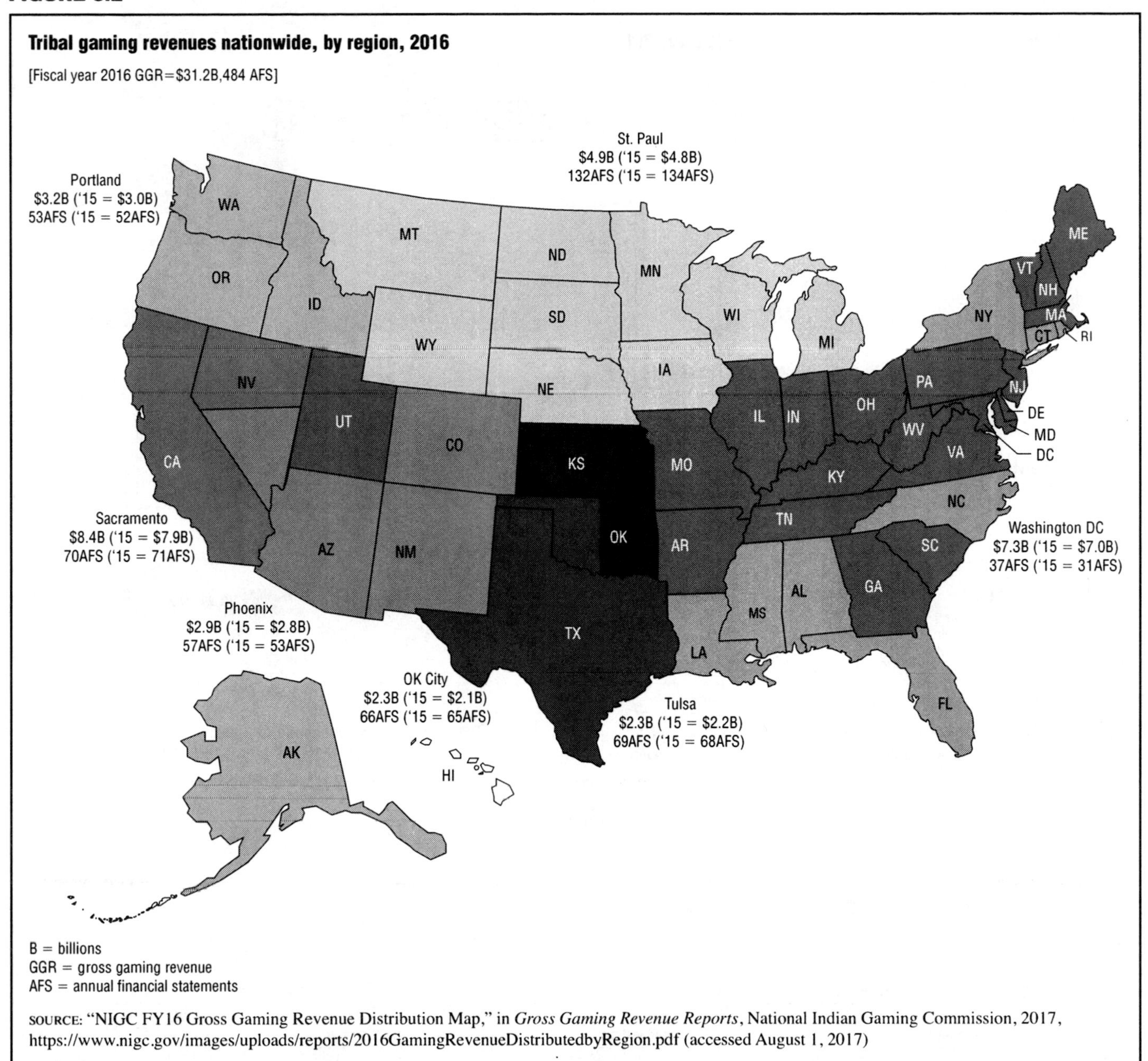

Tribal gaming revenues nationwide, by region, 2016

[Fiscal year 2016 GGR=$31.2B,484 AFS]

B = billions
GGR = gross gaming revenue
AFS = annual financial statements

SOURCE: "NIGC FY16 Gross Gaming Revenue Distribution Map," in *Gross Gaming Revenue Reports*, National Indian Gaming Commission, 2017, https://www.nigc.gov/images/uploads/reports/2016GamingRevenueDistributedbyRegion.pdf (accessed August 1, 2017)

Alabama, Florida, North Carolina, New York, and Connecticut. As of 2016, the St. Paul region was the largest in terms of gaming facilities, with 132 operations spread across a nine-state area. The Sacramento region was the largest in terms of revenues, with $8.4 billion. The Washington, D.C., region closely followed Sacramento in the size of its total revenues, with $7.3 billion. Casinos operating in Connecticut, especially Foxwoods Casino and Resort (located approximately midway between New York City and Boston, Massachusetts, and adjacent to both Providence, Rhode Island, and Hartford, Connecticut), are thought to be the largest source of the region's revenues.

All NIGC regions saw overall revenue growth during both fiscal years (FYs) 2015 and 2016, but there was considerable variation in the rates of growth. (See Figure 5.3.) The region with the highest revenue levels, Sacramento, also experienced the sharpest growth, at 6.3%. The second-highest growth rate occurred in the Oklahoma City region, at 5.7%, followed by Portland (5.1%), Phoenix (4.4%), Tulsa (4%), Washington, D.C. (3.8%), and St. Paul (1.1%). As Table 5.1 shows, the total number of tribal casinos nationwide increased from 474 during FY 2015 to 484 during FY 2016. All but two regions experienced an increase in casinos; the total in St. Paul decreased by two casinos during this span, while the total in Sacramento decreased by one.

The NIGC notes that a majority of tribal gaming facilities are small operations, as measured by revenues.

FIGURE 5.3

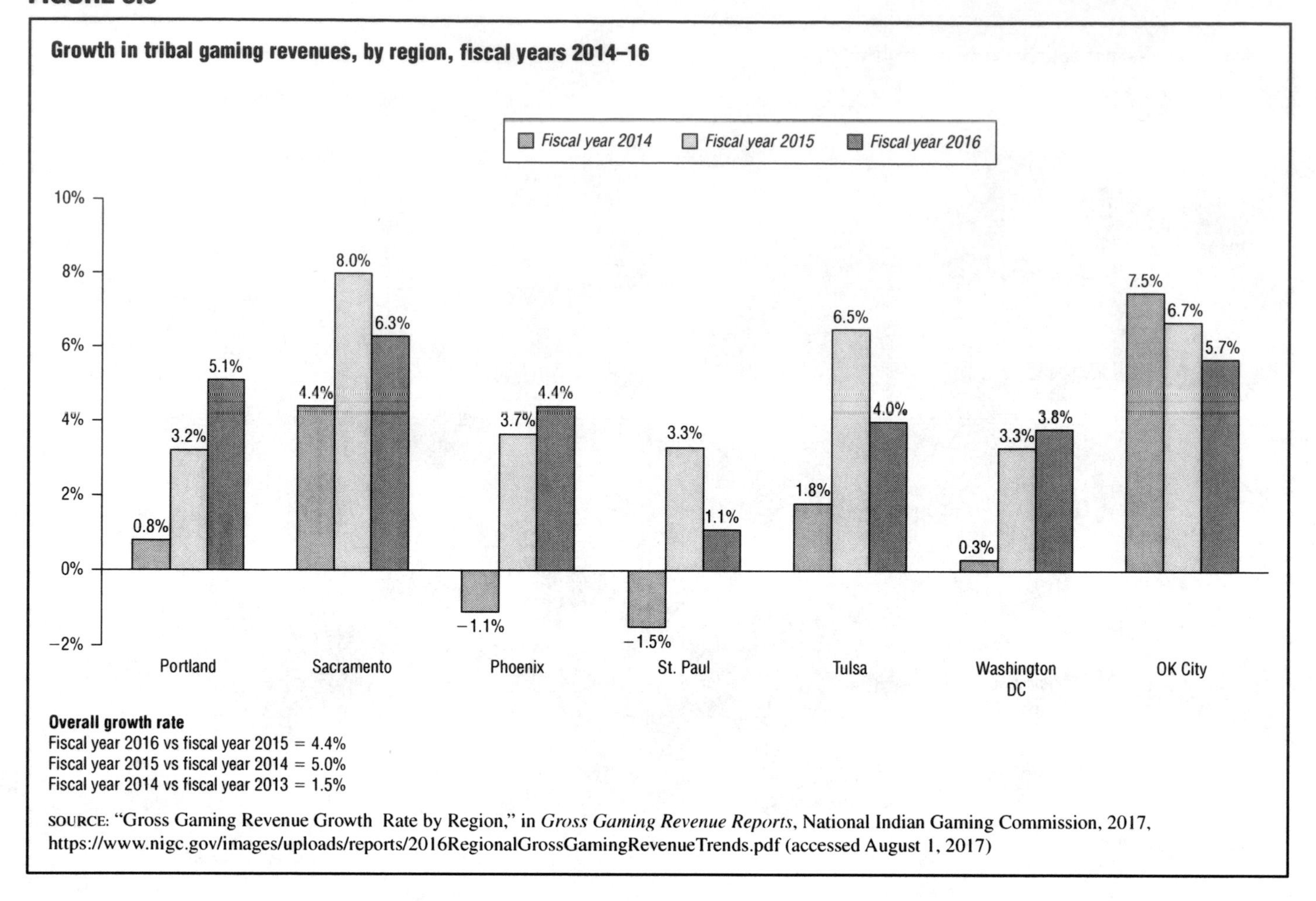

Growth in tribal gaming revenues, by region, fiscal years 2014–16

SOURCE: "Gross Gaming Revenue Growth Rate by Region," in *Gross Gaming Revenue Reports*, National Indian Gaming Commission, 2017, https://www.nigc.gov/images/uploads/reports/2016RegionalGrossGamingRevenueTrends.pdf (accessed August 1, 2017)

TABLE 5.1

Tribal gaming revenues, by region, fiscal years 2015–16

[In thousands]

	Fiscal year 2016		Fiscal year 2015		Increase (decrease)		
	Number of submissions	Gaming revenues	Number of submissions	Gaming revenues	Number of submissions	Gaming revenues	Revenue percentage
Portland	53	$3,177,311	52	$3,022,108	1	$155,203	5.1%
Sacramento	70	$8,380,918	71	$7,881,104	(1)	$499,814	6.3%
Phoenix	57	$2,932,157	53	$2,807,744	4	$124,413	4.4%
St. Paul	132	$4,884,335	134	$4,829,181	(2)	$55,153	1.1%
Tulsa	69	$2,295,202	68	$2,206,538	1	$88,663	4.0%
OK City	66	$2,264,665	65	$2,141,567	1	$123,098	5.7%
Washington DC	37	$7,260,962	31	$6,993,501	6	$267,460	3.8%
Totals	**484**	**$31,195,549**	**474**	**$29,881,744**	**10**	**$1,313,805**	**4.4%**

Portland	Alaska, Idaho, Oregon, and Washington.
Sacramento	California, and Northern Nevada.
Phoenix	Arizona, Colorado, New Mexico, and Southern Nevada.
St. Paul	Iowa, Michigan, Minnesota, Montana, North Dakota, Nebraska, South Dakota, Wisconsin and Wyoming.
Tulsa	Kansas, and Eastern Oklahoma.
OK City	Western Oklahoma and Texas.
Washington DC	Alabama, Connecticut, Florida, Louisiana, Mississippi, North Carolina, and New York.

SOURCE: "Gaming Revenues by Region," in *Gross Gaming Revenue Reports*, National Indian Gaming Commission, 2017, https://www.nigc.gov/images/uploads/reports/2016GamingRevenueDistributedbyRegionChart.pdf (accessed August 1, 2017)

According to the NIGC, in "2016 Indian Gaming Revenues Increase 4.4 Percent," 57% of all tribal gaming revenues in 2016 came from casinos that generated less than $25 million a year. According to standards set by the U.S. government's Small Business Administration, in "Small Business Size Standards: Arts, Entertainment, and Recreation" (June 13, 2013, https://www.regulations.gov/document?D=SBA-2012-0006-0004), all casinos making less than $25.5 million in annual revenues are considered to be small businesses.

The IGRA requires that net revenues from tribal gaming be used to fund tribal government operations or programs, to provide for the general welfare of the tribe and its members, to promote tribal economic development, and to help fund operations of local government agencies. In fact, revenues from tribal gaming often make significant contributions to social services on reservations. For example, in "Native Americans Can't Always Cash In on Casinos" (Guardian.com, August 9, 2010), Barbara Wells indicates that the Muckleshoot Tribe's casino profits have provided seed money for unrelated businesses and funded a medical facility that offers medical and dental care for all tribe members. The profits have also helped the tribe build homes for its elders and provided funds for 200 Washington State charities through the Muckleshoot Charity Fund.

Thaddieus Conner and William Taggart report in "Is the House Winning? Exploring the Impact of Indian Gaming" (2012, https://www.aisc.ucla.edu/gng/posters/thaddiusconner.pdf), the most recent available data on this subject as of August 2017, that tribes spent the net revenues of tribal casinos in the following ways in 2011: 26% was used for infrastructure and housing, 20% was used for education and social welfare programs, 19% was used for economic development, 17% was used for health care, and 17% was used for police and fire protection. Tribes with gaming operations may distribute gaming revenues to individual tribe members through per capita payments but are not required to do so. Such payments must be approved by the U.S. secretary of the interior as part of the tribe's Revenue Allocation Plan and are subject to federal income tax.

TRIBAL-COMMERCIAL CASINO VENTURES

Building casinos can be expensive. Tribes that have built them have had to borrow large sums of money and/or obtain investors to do so. In general, the IGRA requires that tribes partner with companies for no more than five years at a time and limits the companies' take to 30% of the total revenue. Under some circumstances, the partnership can last seven years and the companies' portion can be as much as 40% of the total revenue. These five- to seven-year contracts can also be renewed if both parties and the state government agree to the renewal. In some cases more than one tribe enters into a single casino contract with a private operator, and in other cases one tribe enters into multiple contracts for multiple facilities. According to the NIGC, in "Approved Management Contracts" (https://www.nigc.gov/finance/approved-management-contracts), as of 2017, 59 tribes had 70 management contracts in place with commercial companies.

Native American casinos have often faced fierce opposition from commercial casino operators hoping to thwart competition. For example, tribal casinos in California could cut deeply into the Nevada casino business because California residents, who have long provided a large share of Nevada's gambling revenue, could gamble closer to home. However, some commercial casino operators have seen increased opportunities for revenue through partnerships with Native American tribes, and some tribes—especially small tribes—have welcomed the investment capital and management experience offered by commercial partners.

For example, the Thunder Valley Casino is a Native American gaming venture located about 30 miles (48 km) northeast of Sacramento. The casino, which opened in June 2003, is owned by the United Auburn Indian Community, which has about 300 members. The casino is financially backed and managed by Station Casinos of Las Vegas. The tribe selected the company because it was willing to provide $200 million to build the casino and agreed to manage the casino for the tribe; in return, Station Casinos receives 24% of the Thunder Valley Casino's net revenues. Before the casino opened, tribe members lived in poverty on a 3-acre (1.2-ha) reservation. However, with the casino came full health, dental, and vision insurance for each tribe member. Gaming revenues also funded the United Auburn Indian Community Tribal School, which has a teacher-student ratio of 1 to 7. In June 2010 the tribe opened a new expansion to the casino: a 297-room hotel that had a spa and 14 restaurants and bars. Dale Kasler reports in "Red Hawk Casino's Fortunes Have Disappointed So Far" (SacBee.com, December 11, 2011) that by 2011 each member of the United Auburn Indian Community was receiving an estimated $30,000 per month in profits generated by the Thunder Valley Casino.

Tribal Casinos off the Reservation

Another aspect of tribal gaming is the construction of tribal casinos on land outside of reservations. A common motivation for opening an off-reservation casino involves the location of many tribal reservations. Historically, most reservations were established in remote locations, on land that was not valuable to white settlers or the businesspeople of the time. Although population growth and the development of cities and suburbs have brought large populations nearer to some reservations, many others remain isolated, sometimes by hundreds of miles, from the nearest population center. Tribes that occupy such lands have limited options for economic development, and some have sought to build casinos that are located near population centers to derive the benefits that other, better-situated tribes enjoy. To build a new casino on nonreservation land, tribes must convince the BIA that they have claim to a parcel of land where they would like to build the new casino. The BIA can then put the land into a trust for the tribe.

According to Ian Lovett, in "Tribes Clash as Casinos Move Away from Home" (NYTimes.com, March 3, 2014), this trend has complicated the politics of tribal casinos and raised legal and constitutional issues that remained unresolved as of 2017. According to Lovett, the administration of President George W. Bush (1946–) typically denied applications for casinos located farther than "commuting distance" from a tribe's reservation. In contrast, the administration of President Barack Obama (1961–) took a different view of the issue and was, in general, more willing to approve off-reservation casinos located at greater distances from tribes' reservations. Between 2009 and 2016 the BIA approved four off-reservation casinos: the southern Arizona–based Tohono O'odham Nation's plans for a facility located in the Phoenix suburb of Glendale, 80 miles (129 km) from its reservation; the Menominee Tribe's bid to build a casino in Kenosha, Wisconsin, 45 minutes from Milwaukee but more than 150 miles (240 km) from its reservation; the North Fork Band of Mono Indians' application to build a casino near Madera, California, northwest of Fresno, 36 miles (58 km) from its reservation; and the Spokane Tribe's efforts to open a casino in Airway Heights, Washington, approximately 25 miles (40 km) from the reservation's borders. As of mid-2017, however, only the Tohono O'odham Nation had succeeded in opening an off-reservation casino. During this time the Menominee Tribe and the North Fork Band continued to face resistance from elected officials in their respective states, while the Spokane Tribe did not commence construction of the Airway Heights Casino until April 2017.

Indeed, efforts to build off-reservation casinos, sometimes called "reservation shopping," have generated opposition from a number of different sources. In 2005 the organization Citizens against Reservation Shopping was founded in an attempt to block efforts by the newly recognized Cowlitz Indians to build a casino in Clark County, Washington. The group eventually filed a lawsuit against the federal government, which went before the U.S. Supreme Court in October 2016. The case was initially known as *Citizens against Reservation Shopping v. Haugrud*, after Kevin Haugrud, the acting U.S. secretary of the interior. The suit was subsequently renamed *Citizens against Reservation Shopping v. Zinke*, after the U.S. representative Ryan Zinke (1961–; R-MT) was confirmed as the interior secretary under the newly elected President Donald Trump (1946–). The group's petition was ultimately denied by the court in April 2017. That same month, according to the article "Trump Administration Faces Test of Off-Reservation Gaming Policy" (Indianz .com, April 18, 2017), Zinke announced that subsequent applications for off-reservation casinos would receive approval directly from the Department of the Interior, an indication that Trump intended to assume a harder line against reservation shopping than his predecessor.

Off-reservation casinos have also complicated tribal alliances. Since the advent of tribal gaming during the 1980s, tribes across the United States have typically presented a united front in the face of opposition to the expansion of tribal casinos, but the issue of off-reservation casinos has tended to pit tribes against one another. This is largely due to the fact that tribes wanting to open casinos far from their reservations often threaten to compete with existing tribal casinos. Because commercial and tribal gambling have spread throughout the United States, there are few population centers whose gambling needs are not already being served by either commercial or tribal casinos.

For example, Lovett reports that the North Fork Band's proposed casino outside of Fresno would directly compete with a casino operated by the Chukchansi Indians only 30 miles (48 km) away. Efforts to block the off-reservation casino scored a major victory in 2014, when California voters resoundingly defeated the Referendum on Indian Gaming Compacts (also known as Proposition 48), which would have allowed the North Fork gaming facility to proceed. Meanwhile, another off-reservation casino, run in a small lower Michigan town by the Bay Mills Indian Community, whose reservation is far away in the state's remote Upper Peninsula, led to a legal battle with the state of Michigan. Michigan sued the tribe to close the casino and a federal judge ordered it closed in 2011. The state argued that the tribal sovereignty on which the Bay Mills Indian Community's right to build the casino rested did not apply outside of tribal lands. The case was ultimately decided by the U.S. Supreme Court in May 2014. In *Michigan v. Bay Mills Indian Community* (No. 12-515), the court ruled 5–4 that the tribe, as a sovereign entity, was protected from being sued by the state, effectively dismissing the state's suit and opening the way for the reopening of the Bay Mills casino. Writing for the majority, Justice Elena Kagan (1960–) noted that the state had other options for enforcing what it regarded as the tribe's legal violations, including arresting tribal officials and casino operators, but that it was up to Congress, rather than the courts, to decide issues related to tribal immunity.

The larger issue of whether tribal sovereignty extends beyond the boundaries of a tribe's lands thus remained unresolved. A number of U.S. Senate and U.S. House of Representatives bills over the years have attempted to redefine tribal sovereignty to restrict off-reservation gaming, but none have received the necessary support to pass.

THE STORY OF NATIVE AMERICAN CASINOS IN TWO STATES

Connecticut

Tribal casinos are not required by law to make their financial records public. Although exact figures are not known, various reports indicate that the tribal casinos operating in Connecticut are extremely profitable. As of 2017, only two tribal casinos were operating in Connecticut. Foxwoods Casino and Resort is operated by the Mashantucket Pequot Tribe in Ledyard, and the Mohegan Sun is operated by the Mohegan Tribe in nearby Uncasville. Both are located in a rural area of eastern Connecticut.

In "About Us" (2017, http://www.foxwoods.com/aboutus.aspx), Foxwoods describes itself as being "the premier resort destination in the Northeast." In 2017 it had seven casinos and five hotels, two spas, a golf and country club, a shopping mall, dozens of restaurants, 10 bars and nightclubs, two theaters, and a bowling alley, among other amenities. Foxwoods had more than 4,800 slot machines and more than 250 gaming tables, as well as one of the world's largest bingo halls.

The Mohegan Sun indicates in "About Mohegan Sun" (2017, http://mohegansun.com/about-mohegan-sun.html) that it had 1,563 hotel rooms, more than 40 restaurants and bars, and three casinos in 2017. The complex also included a 10,000-seat arena (home to the Women's National Basketball Association's Connecticut Sun), a live music venue, a comedy club, a shopping mall, a family entertainment facility, a three-story indoor mountain made of crystal, and a 55-foot (17-m) artificial waterfall.

Foxwoods in particular has an interesting history. According to Kim Isaac Eisler, in *Revenge of the Pequots: How a Small Native American Tribe Created the World's Most Profitable Casino* (2002), during the 1980s the Connecticut legislature passed a law that allowed the wagering of "play money" on casino games such as blackjack, roulette, craps, and poker. The Mothers against Drunk Driving organization championed the law to encourage high schools to hold casino-type events following proms to reduce drunk driving by teenagers. Under this law, the Mashantucket Pequot Tribe was able to get a license for a "charity" gambling casino. It also procured $60 million from the resort developer Sol Kerzner (1935–) to begin construction.

Foxwoods opened in 1992. At that time, slot machines were not permitted. In 1994 the tribe negotiated a deal with Lowell P. Weicker (1931–), the governor of Connecticut, that provided the tribe with exclusive rights to operate slot machines within the state. In return, the tribe agreed to make yearly payments to the state of $100 million or 25% of the revenue from the casino's slot machines, whichever was greater. By 1997 Foxwoods was considered to be the largest and most profitable casino in the United States.

In 1994 the Mohegan Tribe also signed a compact with Weicker to operate a casino. The Mashantucket Pequot Tribe granted the Mohegan Tribe permission to include slot machines in its new casino. In return, the state set the annual payment required from each tribe at $80 million or 25% of their slot revenue, whichever was greater. The Mohegan Sun opened in 1996 after receiving financing from Kerzner.

The Mashantucket Pequots' standing as a tribe is not without controversy. In *Without Reservation: How a Controversial Indian Tribe Rose to Power and Built the World's Largest Casino* (2001), Jeff Benedict claims that the Mashantucket Pequots never should have been legally recognized as a tribe by the federal government because some members are not actually descendants of the historic Pequot Tribe. The tribe achieved its recognition by an act of Congress. Benedict made his allegations a major part of his unsuccessful bid for Congress during the summer of 2002. He later helped found the Connecticut Alliance against Casino Expansion, a nonprofit coalition that lobbied against additional casinos in Connecticut. The alliance also sought federal legislation to reform the tribal recognition process.

Legalized gambling in Connecticut is regulated by the Division of Special Revenue, which conducts licensing, permitting, monitoring, and education. It also ensures that the correct revenues are transferred to the state's general fund and to each municipality that hosts a gaming facility or charitable game. The Division of Special Revenue indicates that Connecticut collected $3.8 billion from Foxwoods and $3.2 billion from Mohegan Sun between FYs 1993 and 2015. (See Table 5.2.) These casino revenues represented 42.8% of all gaming payments (including lottery proceeds) to the state general fund between FYs 1993 and 2015.

Nevertheless, the economic benefits of the casinos in Connecticut reach far beyond payments to the state general fund. The Division of Special Revenue analyzes in *Gambling in Connecticut: Analyzing the Economic and Social Impacts* (June 22, 2009, http://www.ct.gov/dosr/lib/dosr/june_24_2009_spectrum_final_final_report_to_the_state_of_connecticut.pdf) the economic impact of Foxwoods and Mohegan Sun and finds that in addition to direct tax revenue (slot machine revenue contributions, regulatory levies, personal income tax, and local property tax), the state also earned approximately $556.4 million in 2007 in indirect taxes, which include personal income taxes from employees in and sales tax from businesses that cater to the casino industry. In addition, Foxwoods

TABLE 5.2

Connecticut tribal gaming payments to state general fund, fiscal years 1993–2015

FYE		Parimutuel							Casino			
6/30	Lottery	Plainfield Greyhound	Bridgeport/ Shoreline Star	Hartford Jai Alai	Milford Jai Alai	Sub-total	Off-track betting	Charitable games	Foxwoods	Mohegan Sun	Subtotal	Grand total
1993	221,700,000	2,578,114	2,632,772	2,962,939	3,138,557	11,312,382	16,200,000	1,735,931	$30,000,000		$30,000,000	280,948,313
1994	217,250,000	682,389	446,604	519,205	713,048	2,361,246	5,788,175	1,805,800	113,000,000		113,000,000	340,205,221
1995	249,650,000	592,446	350,990	421,212	639,706	2,004,354	6,129,150	1,748,657	135,724,017		135,724,017	395,256,178
1996	262,050,000	490,421	210,335	141,034	858,996	1,700,786	6,610,554	1,723,649	148,702,765		148,702,765	420,787,754
1997	251,520,868	308,935	47,231	0	521,138	877,304	6,874,079	1,491,772	145,957,933	$57,643,836	203,601,769	464,365,792
1998	264,274,830	281,153	38,816	0	401,319	721,288	5,441,570	1,423,223	165,067,994	91,007,858	256,075,852	527,936,763
1999	271,308,022	255,094	37,090	0	341,630	633,814	5,472,648	1,258,380	173,581,104	113,450,294	287,031,398	565,704,262
2000	253,598,047	210,483	35,425	0	324,365	570,273	5,616,495	1,205,865	189,235,039	129,750,030	318,985,069	579,975,749
2001	252,002,987	167,740	40,930	0	294,562	503,232	5,674,281	1,162,360	190,683,773	141,734,541	332,418,314	591,761,174
2002	271,509,680	162,945	41,969	0	137,764	342,678	5,736,901	1,284,454	199,038,210	169,915,956	368,954,166	647,827,879
2003	256,814,859	134,743	43,222	0	0	177,965	5,783,231	1,230,391	196,300,528	190,953,944	387,254,472	651,260,918
2004	280,763,074	109,394	43,116	0	0	152,510	5,783,041	1,398,295	196,883,096	205,850,884	402,733,980	690,830,900
2005	268,515,000	64,837	39,462	0	0	104,299	5,275,182	1,431,054	204,953,050	212,884,444	417,837,494	693,163,029
2006	284,864,998	0	25,757	0	0	25,757	5,055,057	1,305,163	204,505,785	223,020,826	427,526,611	718,777,586
2007	279,000,000	0	0	0	0	0	4,808,425	1,297,756	201,380,257	229,095,455	430,475,712	715,581,893
2008	283,000,000	0	0	0	0	0	4,603,607	1,211,178	190,037,675	221,373,298	411,410,973	700,225,758
2009	283,000,000	0	0	0	0	0	4,195,243	1,063,435	177,153,485	200,651,400	377,804,885	666,063,563
2010	285,500,000	0	0	0	0	0	3,816,676	945,375	169,408,149	189,845,097	359,253,246	649,515,297
2011	289,300,161	0	0	0	0	0	3,699,415	876,064	174,092,415	185,488,712	359,581,127	653,456,767
2012	310,000,000	0	0	0	0	0	3,766,758	683,627	165,547,090	178,783,321	344,330,411	658,780,796
2013	312,100,270	0	0	0	0	0	3,644,167	425,956	138,531,943	157,863,949	296,395,892	612,566,285
2014	319,500,000	0	0	0	0	0	3,723,791	416,962	131,527,642	148,345,903	279,873,545	603,514,298
2015	319,700,000	0	0	0	0	0	3,606,125	415,929	121,302,985	146,682,864	267,985,849	591,707,903
	$8,529,784,716	$141,970,035	$77,420,591	$76,296,801	$75,884,044	$371,571,471	$374,804,571	$33,528,962	$3,762,614,935	$3,194,342,612	$6,956,957,547	$16,266,647,267

FYE = fiscal year ended.
OTB = off-track betting.
Notes:
1. Revenue transferred on cash basis per fiscal year.
2. The above transfers represent:
a) actual Lottery transfers through June 30, 2015 as reported by the Connecticut Lottery Corporation.
b) collection of parimutuel taxes, net of payments to municipalities and other entities, for the former jai alai and greyhound facilities.
c) collection of parimutuel taxes, net of payments to municipalities and other entities, for races conducted through June 30, 2015 for off-track betting.
d) Sealed ticket and bingo revenue through June 30, 2015.
e) actual casino contributions through July 15, 2015, based on reported video facsimile/slot machine revenue through June 30, 2015.
3. From its inception in 1976 through June 30, 1993, the off-track betting (OTB) system was state operated. For that period, transfers represented the fund balance in excess of division needs. The OTB system was sold to a private operator effective July 1, 1993 and since then transfers are based on a statutory parimutuel tax rate.

SOURCE: Adapted from "Transfers to General Fund," in *Gaming Revenues and Statistics*, Connecticut Department of Consumer Protection, April 1, 2016, http://www.ct.gov/dcp/lib/dcp/pdf/gaming/stmt2015.pdf (accessed June 28, 2017)

and Mohegan Sun generated a combined 32,510 direct, indirect, and induced jobs in Connecticut in 2007. The total gross regional product contributed by the two casinos was estimated at $1.9 billion in 2007, and personal income generation for state residents was estimated at $1.2 billion.

California

The Judicial Branch of California notes in "California Tribal Communities" (http://www.courts.ca.gov/3066.htm) that California had 109 federally recognized tribes in 2017. Most are small extended family groups living on a few acres of federal trust property called *rancherias*. Some tribes have only a handful of members.

According to the AGA, in *2016 State of the States*, there were 71 tribal casinos in California in 2015. This total is at odds with the NIGC information included in Figure 5.2, which indicates that in 2015 there were 71 tribal gambling facilities in the Sacramento region, which consists of California and northern Nevada. In any event, there is no contesting the fact that California has more tribal casinos than any other state and that the Sacramento region's $8.4 billion in 2016 revenues represented more than a quarter (26.9%) of all tribal casino revenues nationally.

Before 2000 the gambling operations of California tribes were largely limited to bingo halls because state law prohibited the operation of slot machines and other gambling devices, certain card games, banked games, and games where the house collects a share of the amount wagered. In 2000 California voters passed Proposition 1A, which amended the state constitution to permit Native American tribes to operate lottery games, slot machines, and banking and percentage card games on tribal lands. The constitutionality of the measure was immediately challenged in court.

In January 2002 Gray Davis (1942–), the governor of California, signed 62 gambling compacts with California tribes. The compacts allowed each tribe to have a maximum of 2,000 slot machines. The governor also announced plans to cap the number of slot machines in the state at 45,000. At the time, there were already 40,000 slot machines in operation and dozens of tribal casinos in the planning stages. The governor put a moratorium on new compacts while Proposition 1A made its way through the courts. In August 2002 a U.S. district court ruled that tribal casinos were entitled to operate under the provisions of the state gaming compacts and Proposition 1A.

In 2003 the state of California suffered a severe budget crisis. Davis was ultimately forced out of office through a special recall election in which Arnold Schwarzenegger (1947–) became the governor. Having campaigned on promises to extract more state government revenues from tribal casinos, Schwarzenegger signed in June 2004 new compacts that preserved the exclusive gaming rights of five California tribes: the Pala Band of Mission Indians, the Pauma Band of Mission Indians, the Rumsey Band of Wintun Indians, the United Auburn Indian Community, and the Viejas Band of Kumeyaay Indians. The slot machine cap was also raised above 2,000 machines per tribe. In exchange, the tribes agreed to pay the state $1 billion up front and a licensing fee for each new slot machine that was added above the current limit. Payments were expected to total between $150 million and $275 million per year through 2030, when the compacts are set to expire.

The state continued to form tribal compacts that permitted casino expansion in exchange for higher taxes. One of the more notable deals was made with the Agua Caliente Band of Cahuilla Indians. The tribe owned two casinos in Palm Springs. In August 2006 the state agreed to let the tribe open a third casino with 5,000 slot machines if the tribe paid an estimated $1.9 billion in taxes over the following 23 years. Many Californians were concerned that if such deal making were allowed to continue, casinos could be as prevalent as shopping malls and lead to higher instances of gambling addiction.

In 2007 four referenda petitions were proposed to overturn compacts that had recently been negotiated between four tribes—the Agua Caliente Band of Cahuilla Indians, the Morongo Band of Mission Indians, the Pechanga Band of Luiseño Indians, and the Sycuan Band of the Kumeyaay Nation—and the state. Although the compacts had already been signed by Schwarzenegger and ratified by the legislature, a petition campaign that was supported by the California Federation of Teachers, the American Indian Rights and Resource Organization, the California Tax Report Association, and the United Farm Workers succeeded in getting approval of the new compacts on the ballot in February 2008. California voters were asked to approve or disapprove of the four new compacts with a "yes" or "no" vote on Propositions 94, 95, 96, and 97. Collectively, the compacts allowed the four tribes to operate an extra 17,000 slot machines with an additional payment of $9 million per year to the Revenue Sharing Trust Fund. Voters approved the four compacts. In October 2009 the Ninth Circuit Court of Appeals in California had cleared the way for the issuing of 3,547 new gaming device licenses to 11 other tribes in the state.

Onell R. Soto notes in "Rincon Tribe Wins Slot Suit against State" (UTSanDiego.com, March 27, 2010) that in March 2010 the Rincon Band of Luiseño Indians won its lawsuit against the state over the number of slot machines it could operate at its casino in North County. A federal judge ruled that the state should permit 56,000 slot machines under compacts signed with Governor Davis in

1999. However, because some tribes had renegotiated with California and received additional slot machines in exchange for a larger payment to the state, there were already approximately 65,000 machines in tribal casinos throughout California in March 2010.

In *2016 California Tribal Gaming Impact Study: An Economic, Fiscal, and Social Impact Analysis with Community Attitudes Survey Assessment* (August 2016, http://cniga.com/wp-content/uploads/2017/02/12-2-16-Full-Report-Final.pdf), the California Nations Indian Gaming Association provides data on the economic impact of tribal gaming in the state. According to the association's findings, tribal gaming directly or indirectly injected $7.8 billion in California's economy and accounted for 63,000 jobs in 2014. This included spending by the tribes themselves (primarily on advertising, administration, food and beverages, and gaming operations) and secondary economic effects (e.g., when tribal members who benefit directly from casinos go on to purchase goods and services from other businesses not directly involved in gaming). Overall, tribal gaming generated $392.4 million in state and local tax revenues in 2014.

CHAPTER 6

THE ECONOMIC AND SOCIAL EFFECTS OF CASINOS

Assessing the effects of casinos on society is complicated because many factors have to be considered. The economic effects of casinos are difficult to capture precisely and are subject to dispute among economists, and separating the social effects of casinos from subjective judgments regarding quality of life and morality is similarly difficult.

Proponents of the gambling industry consider casinos to be part of the leisure and entertainment sector—like amusement parks or movie theaters, in which participants exchange their money for a good time. They generally do not frame gambling in moral terms but point to its financial impact on local, state, and national economies. Opponents provide a variety of reasons for their disapproval of casino gambling. Some doubt the economic effects of casinos and argue that other economic development strategies are preferable. Others disapprove of gambling on religious grounds or are wary of an industry that has historically been associated with mobsters, swindlers, and corrupt politicians. Still others point out that casinos provide a place for those who are prone to problem gambling to act on those urges. Easy accessibility to casinos, they suggest, encourages some people to gamble who otherwise would not and should not. Problem gamblers impose significant costs on their families, communities, and the health care system.

NATIONAL PUBLIC OPINION

As Figure 1.1 in Chapter 1 shows, 65% of Americans considered gambling morally acceptable in 2017. According to the American Gaming Association (AGA), in *2013 State of the States: The AGA Survey of Casino Entertainment* (2013, https://www.americangaming.org/sites/default/files/research_files/aga_sos2013_rev042014.pdf), 47% of a representative sample of Americans in 2013 considered casino gambling acceptable for anyone, 38% considered it acceptable for others but not for themselves, and 14% considered it universally unacceptable. No unbiased recent polling offering a more detailed sense of Americans' views on gambling was available as of August 2017, but a 2006 report issued by the Pew Research Center's Paul Taylor, Cary Funk, and Peyton Craighill indicates that Americans at that time had mixed feelings about the spread of casino gambling to virtually all parts of the United States. The researchers note in *Gambling: As the Take Rises, So Does Public Concern* (May 23, 2006, http://assets.pewresearch.org/wp-content/uploads/sites/3/2010/10/Gambling.pdf) that 42% of those surveyed said casinos are detrimental to their communities, whereas 34% said casinos have a positive impact. A smaller percentage of people who lived near a casino (38%) had a negative view of the casino's influence on their community than those who did not live near a casino (45%).

In *Economic and Social Impact of Introducing Casino Gambling: A Review and Assessment of the Literature* (March 2010, http://www.philadelphiafed.org/community-development/publications/discussion-papers/discussion-paper_casino-gambling.pdf), Alan Mallach of the Brookings Institution notes that the growing prevalence of casinos since the 1990s does not necessarily suggest that growing numbers of people fully approve of the industry or the opening of casinos near their community. According to Mallach, the ambivalence that many Americans feel about casinos takes the form of public approval of the restrictive legal, regulatory, and tax structures under which the commercial gaming industry operates. Although early entrants to the casino industry, such as Nevada and New Jersey, tax casinos at rates similar to other businesses, more recent state laws that allow for casino gambling have included extensive regulation schemes and tax rates well in excess of those paid by other types of businesses, in some cases exceeding 50%. Casinos win public approval in spite of significant public anxiety about their effects on society, Mallach suggests, because of the perception that they offer prospects for economic development that are not easily obtainable by other means.

Additionally, the Spectrum Gaming Group observes in *Gambling Impact Study* (October 28, 2013, http://www.leg.state.fl.us/GamingStudy/docs/FGIS_Spectrum_28Oct2013.pdf) that the process whereby voters decide whether to legalize casinos strongly favors proponents of gambling: "Those who oppose gaming's expansion often succeed, but in many instances they have to continue doing battle in subsequent years. They have to win every time. Those who favor the expansion of gaming need to win only once." For example, between 1990 and 2008, four separate ballot measures proposing to legalize casino gambling in Ohio were voted down. A fifth measure in 2009 passed with a slim majority of votes (53%) at the height of the Great Recession, which lasted from late 2007 to mid-2009. As many analysts have pointed out, voters are most likely to approve of legalizing casinos during recessions and other periods of economic anxiety.

ECONOMIC EFFECTS

The commercial casino industry had revenues of $38.5 billion in 2015, as the AGA notes in *2016 State of the States: The AGA Survey of the Casino Industry* (November 2016, https://www.americangaming.org/sites/default/files/2016%20State%20of%20the%20States_FINAL.pdf). That same year tribal casinos had collective revenues of $29.9 billion, according to the National Indian Gaming Commission in "Gross Gaming Revenue Trending" (July 16, 2017, https://www.nigc.gov/images/uploads/reports/2016GrossGamingRevenueTrends.pdf). Thus, legal casinos in the United States took in $68.4 billion in gaming revenues in 2015, an amount that does not include the substantial nongaming proceeds (from lodging, food and beverage, entertainment, and other goods and services) enjoyed by the tribes and companies that own and operate casinos.

Indeed, especially at the largest resort-style casinos in Las Vegas, gaming revenues have accounted for a decreasing share of overall revenues since the early 1990s. According to the Spectrum Gaming Group, in *Gambling Impact Study*, in 1992 casinos in Clark County, Nevada (which includes Las Vegas and the surrounding smaller gaming markets), derived 61% of their revenues from gaming. Two decades later gaming accounted for only 43% of revenues, whereas lodging, food and beverage, and other forms of nongaming revenues had all risen substantially.

In 2015 the tribal and commercial casino industries' gaming revenues alone were nearly six times higher than the U.S. film industry's North American box office receipts, which were $11.1 billion, according to the Motion Picture Association of America in *Theatrical Market Statistics, 2015* (April 2016, http://www.mpaa.org/wp-content/uploads/2016/04/MPAA-Theatrical-Market-Statistics-2015_Final.pdf). U.S. gaming revenues were also higher than the $63.9 billion in revenues generated that year by the entire North American sports industry, as reported by Darren Heitner in "North American Sports Market at $75.7 Billion by 2020, Led by Media Rights" (Forbes.com, October 10, 2016), an amount that includes sales of tickets, media rights, sponsorships, and merchandise at the professional, college, and minor league levels.

Most commercial casinos have been major successes for their investors, who range from middle-class stockholders to billionaires such as Steve Wynn (1942–) and Sheldon Adelson (1933–). Most tribal casinos have been economically successful as well, bringing wealth and quality-of-life improvements to Native American reservations, many of which had never seen successful economic development of any kind. The Spectrum Gaming Group reports in *Gambling Impact Study* that casinos have no raw material, inventory, or research and development costs. Accordingly, they tend to be efficient business operations with high profit margins. As measured by earnings before interest, taxes, depreciation, and amortization (a commonly used metric for measuring a business's profitability), both commercial and tribal casino companies enjoy profit levels that compare favorably with other major players in the leisure and entertainment sector, such as publicly traded cruise-ship operators and hotel owners.

Casinos do not simply make money for their owners, however. They employ hundreds of thousands of people, and these people in turn support their families, pay taxes, and buy goods and services—factors that contribute to the economic health of their communities, states, and the nation as a whole. Additionally, tribal and commercial casino owners spend large sums of money on their operations (e.g., on equipment purchases, construction and maintenance, advertising, and food and beverage supplies, among many other types of spending) that increase the incomes of other businesses and, in turn, fuel further tax revenues. Tribal casinos fund tribal governments, and although they do not pay federal or state taxes, they pay billions of dollars annually to compensate states and municipalities for regulatory and public service expenses. Commercial casinos are generally taxed by state and local governments at much higher rates than other industries, making them key funders of many important services. Although governments incur increased costs for more police, roads, and sewers when commercial casinos open, casino taxes and fees help fund programs that improve the quality of life in the immediate vicinity or state.

However, even though casinos provide vital employment opportunities and tax revenues, legalized gambling has also been tied to negative economic effects. David Frum observes in "A Good Way to Wreck a Local Economy: Build Casinos" (TheAtlantic.com, August 7,

2014) that when casinos become the center of economic activity in a community, other local businesses struggle to attract customers. "A casino is not like a movie theater or a sports stadium, offering a time-limited amusement," Frum explains. "It is designed to be an all-absorbing environment that does not release its customers until they have exhausted their money."

Government Revenue

In *2016 State of the States*, the AGA reports that commercial casinos generated tax revenues of $8.9 billion in 2015, a 2.9% increase over tax revenues of $8.6 billion in 2014. Pennsylvania generated the most gambling tax revenue in 2015 ($1.4 billion), followed by Nevada ($889.1 million), New York ($888.4 million), Louisiana ($632.2 million), Indiana ($608.1 million), and Ohio ($545.4 million). As the AGA notes, the spread of commercial casinos beyond Nevada and New Jersey has been accompanied by a general tendency toward increasingly higher state tax rates over time. Thus, Pennsylvania, where casino gambling was legalized in 2006, enjoyed 2015 gaming tax contributions that were more than one-third higher than those of Nevada, despite having casino revenues that were more than three times smaller. Pennsylvania's casino revenues were $3.2 billion, and Nevada's were $11.1 billion in 2015.

The revenues earned by tribal casinos are not taxable because the casinos are operated by tribal governments. Just as the U.S. government does not tax the states for revenue earned from lottery tickets, it does not tax tribal governments for revenue earned from casinos. Likewise, as sovereign entities that exist on a government-to-government basis with the United States, tribes are not subject to state taxes. Tribal members who live on reservations and are employed at tribal enterprises, such as casinos, are not subject to state income taxes. However, members do pay federal income tax, Federal Insurance Contributions Act tax, and Social Security tax on their wages, even if those wages are earned at tribal enterprises. Wages paid to nontribal employees (who account for the majority of employees at many tribal casinos) are subject to state and federal income taxes.

Accordingly, tribal casinos contribute less to state and local governments than do commercial casinos, and many of their contributions are indirect. Reliable current statistics about tribal casino tax contributions, like other statistics relating to the operation of tribal casinos, are for the most part not made public. Nevertheless, the California Nations Indian Gaming Association indicates in *2016 California Tribal Gaming Impact Study: An Economic, Fiscal, and Social Impact Analysis with Community Attitudes Survey Assessment* (August 2016, http://cniga.com/wp-content/uploads/2017/02/12-2-16-Full-Report-Final.pdf) that tribal gaming operations generated an estimated $7.8 billion in economic output that made its way back into the California economy in 2014. As a result, although the tribes that profited from casinos in the state do not pay state and local taxes, their gaming operations generated economic activity that resulted in $392.4 million in state and local taxes.

Most debates about the legalization of gambling have involved disagreements about whether the state tax revenues generated by casinos outweigh the social costs of making gambling readily available in a new locality. These debates have largely assumed that the net effect of casino gambling on government revenues is always a positive one, given that the arrival of casinos in a new location represents the creation of a new source of tax revenue. However, in *Casinonomics: The Socioeconomic Impacts of the Casino Industry* (2013), Douglas M. Walker of the College of Charleston casts doubt on this assumption. Using data from 1985 to 2000 in an attempt to quantify the size of casinos' effect on state tax revenues, Walker finds that "the existence of casinos in a state is associated with a decrease of net state revenue of $90 million."

This surprising result is explained by the fact that gambling's contribution to state tax revenues depends on a number of variables, including tax rates, the size of the gaming industry in a given state, and the industry's relationship to other industries (both gaming and nongaming). Rather than representing an entirely new and separate source of government funding, casino gambling represents consumer spending that would have been directed elsewhere in the absence of casinos, and depending on the size of the gaming industry and the rate of taxation, more state revenues might be generated through other means. It is important to note that Walker's calculations concern a theoretical "average" gaming state rather than, for example, Nevada. Because Nevada's gaming industry is so large and so central to the state's economy, it is hardly likely that the state as it currently exists would enjoy greater tax revenues in the absence of the commercial casino industry. Additionally, Walker notes that, given the size of total state revenues, the $90 million negative average figure is not a statistically significant decrease. In any event, the results of Walker's calculations, should they be validated by further study, undercut one of the primary arguments for the existence of commercial casinos.

EMPLOYMENT AND CAREERS

According to the AGA, in *2016 State of the States*, 350,824 people were employed in commercial casinos in 2015. Nevada was far and away the state with the most commercial casino jobs in 2015, with 170,618 casino employees, or 48.6% of the total commercial casino jobs in the United States. The state with the second-largest population of casino employees, New Jersey, supported 23,615 commercial gaming jobs. Mississippi (20,932 casino employees), Louisiana (19,707), Pennsylvania (17,617), Indiana (14,524), and Colorado (10,775) all had more than 10,000 residents who depended on commercial casinos for employment, and an additional 14 states had between 1,000 and 10,000 casino employees. Gale Courey Toensing reports in "Latest Gaming Industry Report: Indian Gaming Made Small Gains in 2011" (IndianCountryMediaNetwork.com, March 26, 2013) that tribal casinos employed approximately 339,000 people in 2011, the most recent year for which data were available.

The AGA indicates that commercial casinos paid $14.4 billion in wages in 2015, up from $13.2 billion in 2012. According to the U.S. Department of Labor, the median annual wage (the middle value; half made less money and half made more money) for individuals of all occupations in the United States was $37,040 in May 2016, and the median annual wage for individuals employed in the gaming services industry was $20,810. (See Figure 6.1.) All gaming services occupations require extensive customer-service skills and rigorous attention to rules and financial calculations. Supervisory and managerial positions, not surprisingly, pay much better than positions primarily involving direct interaction with customers. (See Table 6.1.) Median annual salaries range from $19,290 for table-game dealers to $69,180 for gaming managers.

All casino employees—from managers to dealers to slot repair technicians—must be at least 21 years old and have licenses from the appropriate regulatory agency. Obtaining a license requires a background investigation—applicants can be disqualified from casino employment for a variety of reasons, including links to organized crime, a felony record, and gambling-related offenses. Requirements for education, training, and experience are up to individual casinos. The U.S. Bureau of Labor Statistics (2016, https://www.bls.gov/ooh/personal-care-and-service/gaming-services-occupations.htm) considers the overall

FIGURE 6.1

Median annual wages, gaming service occupations vs. all occupations, May 2016

[Median annual wages, May 2016]

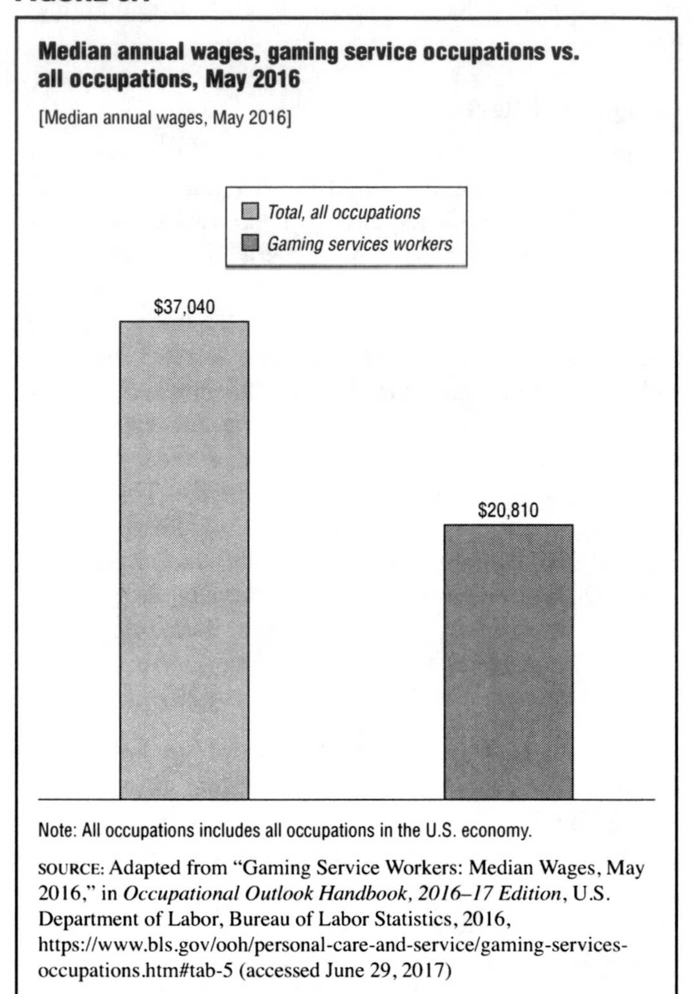

Note: All occupations includes all occupations in the U.S. economy.

SOURCE: Adapted from "Gaming Service Workers: Median Wages, May 2016," in *Occupational Outlook Handbook, 2016–17 Edition*, U.S. Department of Labor, Bureau of Labor Statistics, 2016, https://www.bls.gov/ooh/personal-care-and-service/gaming-services-occupations.htm#tab-5 (accessed June 29, 2017)

TABLE 6.1

Casino occupations, 2016

Title	Responsibilities	Median salary
Gaming managers and supervisors	Direct and oversee the gaming operations and personnel in their assigned area. Supervisors circulate among the tables to make sure that everything is running smoothly and that all areas are properly staffed.	$69,180 for managers; $50,520 for supervisors
Slot supervisors	Oversee the activities of the slot department. The job duties of this occupation have changed significantly, as slot machines have become more automated in recent years. Because most casinos use video slot machines that give out tickets instead of cash and thus require very little oversight, workers in this occupation spend most of their time providing customer service to slot players.	$36,080
Gaming and sports book writers and runners	Handle bets on sporting events and take and record bets for customers. Sports book writers and runners also verify tickets and pay out winning tickets. In addition, they help run games such as bingo and keno. Some gaming runners collect winning tickets from customers in a casino.	$22,600
Gaming dealers	Operate table games such as craps, blackjack, and roulette. They stand or sit behind tables while serving customers. Dealers control the pace and action of the game. They announce each player's move to the rest of the table and let players know when it is their turn. Most dealers are often required to work at least two games, usually blackjack or craps.	$19,290

SOURCE: Adapted from "What Gaming Services Workers Do," in *Occupational Outlook Handbook, 2016–17 Edition*, U.S. Department of Labor, Bureau of Labor Statistics, 2016, https://www.bls.gov/ooh/personal-care-and-service/gaming-services-occupations.htm#tab-2 (accessed June 29, 2017)

FIGURE 6.2

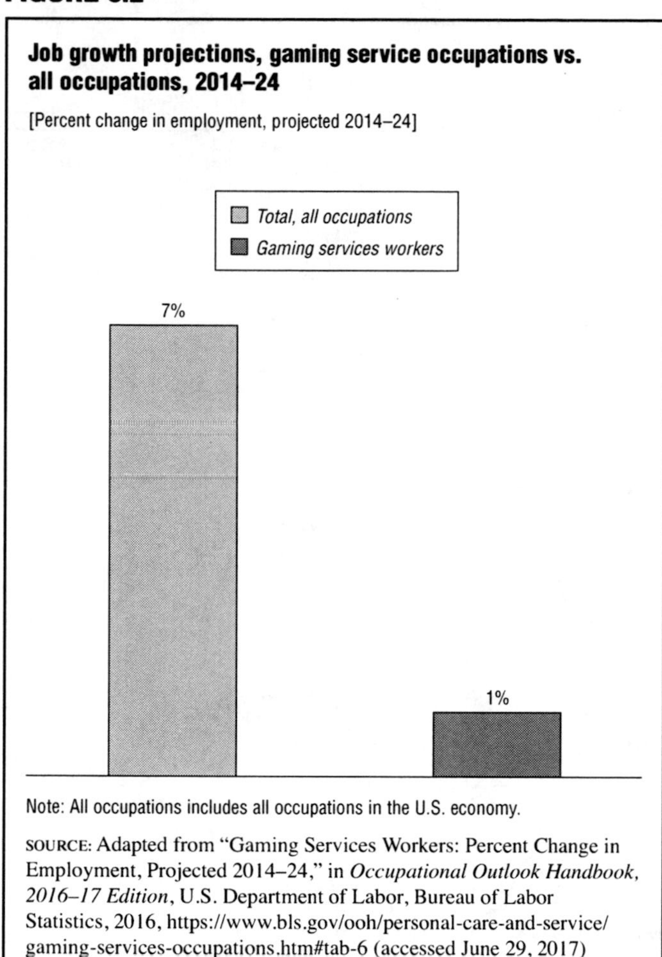

Job growth projections, gaming service occupations vs. all occupations, 2014–24

[Percent change in employment, projected 2014–24]

Note: All occupations includes all occupations in the U.S. economy.

SOURCE: Adapted from "Gaming Services Workers: Percent Change in Employment, Projected 2014–24," in *Occupational Outlook Handbook, 2016–17 Edition*, U.S. Department of Labor, Bureau of Labor Statistics, 2016, https://www.bls.gov/ooh/personal-care-and-service/gaming-services-occupations.htm#tab-6 (accessed June 29, 2017)

employment outlook for the industry to be substantially lower than that for all occupations, with gaming services job opportunities projected to grow at a rate of only 1% between 2014 and 2024, compared with 7% for all occupations. (See Figure 6.2.)

Table 6.2 provides a breakdown of gaming service jobs in Nevada casinos between 1990 and 2016. The number of gaming service jobs peaked in 2001, when 59,908 people worked in gaming service jobs in Nevada. This figure fluctuated between 2002 and 2008, before entering a steady decline, falling to 40,922 in 2016. In *Nevada Statewide Casino Employment—Productivity, Revenues, and Payrolls: A Statistical Study, 1990–2016* (2017, http://gaming.unlv.edu/reports/nvst_emp.pdf), David G. Schwartz and Alexis Rajnoor of the Center for Gaming Research at the University of Nevada, Las Vegas, note that the total number of people employed by Nevada casinos in 2016 was 166,741. Therefore, gaming service jobs accounted for less than a quarter (24.5%) of all Nevada casino jobs that year. By contrast, the total number of people employed by Louisiana riverboat casinos was 14,341 in fiscal year (FY) 2015–16. (See Table 6.3.)

Proponents of casinos typically seek to garner public support for the industry by noting the purported positive effects on job creation and average wages that result from the legalization of gambling. Because the opening of a new casino, especially in areas where casinos did not previously exist, will tend to increase the demand for labor, basic economic theory suggests that the price of labor (the wages people in an area can expect to receive) will rise. Casino backers also note that in many cases gaming jobs pay more than comparable food-service and hospitality jobs. However, critics contend that casinos offer only low-paying jobs without possibilities for career advancement and that casinos derive some portion of their revenues through cannibalizing the sales of preexisting leisure and entertainment businesses (such as hotels and restaurants), thus mitigating at least in part the expected increase in demand for labor.

Among the most reputable studies analyzing the labor market effects of casinos is Chad D. Cotti's "The Effect of Casinos on Local Labor Markets: A County Level Analysis" (*Journal of Gambling Business and Economics*, vol. 2, no. 2, 2008). Cotti's analysis of county-level data in counties where casinos had recently opened suggests that casinos generate modest overall increases in employment and wages. The employment and wage increases are highest in areas with a low population and lowest in areas with a high population.

Cotti's findings are consistent with some of the industry's most widely cited success stories, such as that of Tunica County, Mississippi. Walker notes in *Casinonomics* that prior to the opening of casinos in 1992, rural Tunica County was widely considered to be the poorest county in the United States, with an unemployment rate of 26% and no significant industry other than agriculture. By 1994 the county's unemployment rate had fallen to 4.9%, the number of welfare recipients had decreased 42%, and the number of food stamp recipients had declined 13%. Meanwhile, land prices in some parts of the county had soared by a factor of 100, with parcels previously valued at $250 per acre rising in value to $25,000 per acre. Cotti's contention that casinos bring benefits in employment and wages in inverse proportion to an area's population size is also consistent with one of the casino industry's most glaring disappointments: the failure to spur sustained economic development in Atlantic City, New Jersey. More than three decades after casinos opened in the densely populated area, Atlantic City's overall economic outlook remained bleak by most accounts as of 2017.

The lack of publicly available tribal casino data makes the employment effects of tribal casinos harder to isolate. However, Cotti's observed relation between employment/wages and population is consistent with the overall success of tribal casinos (which varies from tribe to tribe and casino to casino) in promoting economic development. Because many reservations are sparsely

TABLE 6.2

Nevada casino industry employment data, 1990–2016

Title	Employees	Revenue	Rev/Emp	Payroll	Pay/Emp
1990	52,307	4,952,978,992	94,690.56	1,026,029,649	19,616
1991	52,324	5,513,297,150	105,368.42	1,109,535,331	21,205
1992	53,052	5,584,559,864	105,265.77	1,118,540,900	21,084
1993	49,544	5,880,592,442	118,694.34	1,133,598,917	22,881
1994	54,707	6,504,348,451	118,894.26	1,245,500,913	22,767
1995	54,338	7,030,994,404	129,393.69	1,337,572,041	24,616
1996	58,603	7,390,435,180	126,110.19	1,400,234,400	23,894
1997	57,908	7,493,752,537	129,407.90	1,433,142,163	24,749
1998	55,614	7,743,934,793	139,244.34	1,493,595,897	26,856
1999	57,759	8,426,224,115	145,885.91	1,576,888,232	27,301
2000	58,474	9,308,916,419	159,197.53	1,573,781,139	26,914
2001	59,908	9,310,594,712	155,414.88	1,784,980,685	29,795
2002	54,144	8,911,540,280	164,589.62	1,720,207,297	31,771
2003	52,036	9,250,496,837	177,771.10	1,723,704,688	33,125
2004	51,437	9,883,510,737	192,147.88	1,746,423,365	33,953
2005	52,529	10,662,454,467	202,982.25	1,761,628,575	33,536
2006	55,644	11,809,095,031	212,225.85	1,846,166,424	33,178
2007	52,341	12,480,790,793	238,451.52	1,908,249,204	36,458
2008	52,807	12,040,879,888	228,016.74	1,917,906,672	36,319
2009	46,322	10,514,718,393	226,991.89	1,753,571,644	37,856
2010	44,470	9,906,558,446	222,769.47	1,676,694,411	37,704
2011	43,846	10,168,621,415	231,916.74	1,694,893,796	38,656
2012	43,454	10,283,752,257	236,658.36	1,734,534,553	39,917
2013	42,357	10,395,664,377	245,429.67	1,704,473,692	40,241
2014	42,225	10,641,153,886	252,010.75	1,690,639,167	40,039
2015	42,515	10,622,398,618	249,850.61	1,727,351,144	40,629
2016	40,922	10,760,756,016	262,957.72	1,717,095,743	41,960
%Δ	−21.77%	117.26%	177.70%	67.35%	113.91%

Rev/Emp = Revenue divided by employees; used as a measure of productivity.
Pay/Emp = Payroll divided by employees; measures how much, on average, each employee costs.
%Δ = Percent change from 1990 to the present.

SOURCE: Adapted from David G. Schwartz and Alexis Rajnoor, "Statewide Casinos: Casino Department," in *Nevada Statewide Casino Employment—Productivity, Revenues, and Payrolls: A Statistical Study, 1990–2016*, University of Nevada, Las Vegas, Center for Gaming Research, 2017, http://gaming.unlv.edu/reports/nvst_emp.pdf (accessed June 29, 2017)

TABLE 6.3

Louisiana riverboat casino employment data, by location, fiscal year 2015–16

			Total employment			
Licensee	**D/B/A and location**	**Total payroll**	**Total employed[a]**	**Louisiana residents[a]**	**Minorities[a]**	**Female[a]**
Belle of Orleans, LLC	Amelia Bella Casino—Amelia	$7,715,171	333	331	166	214
Catfish Queen Partnership in Commendam	Belle of Baton Rouge—Baton Rouge	19,257,997	539	532	438	321
Eldorado Casino Shreveport Joint Venture	Eldorado Resort Casino Shreveport—Shreveport	23,791,268	1,136	1,115	830	710
Horseshoe Entertainment, L.P.	Horseshoe Bossier City Casino & Hotel—Bossier City	38,288,642	1,287	1,101	743	624
Louisiana Casino Cruises, Inc.	Hollywood Casino Baton Rouge—Baton Rouge	14,558,535	419	417	294	215
Louisiana I Gaming L.P.	Boomtown New Orleans—New Orleans	15,773,901	655	658	462	366
Louisiana Riverboat Gaming Partnership	DiamondJacks Casino & Resort—Bossier City	15,600,020	495	490	381	295
Golden Nugget Lake Charles, LLC[b]	Lake Charles	69,331,592	2,127	2,001	988	1,111
PNK (Baton Rouge) Partnership	L'Auberge Casino & Hotel Baton Rouge—Baton Rouge	21,160,374	1,074	1,054	688	536
PNK (Bossier City), Inc.	Boomtown Bossier City—Bossier City	12,919,561	553	547	387	339
PNK (Lake Charles), LLC	L'Auberge Lake Charles—Lake Charles	73,297,258	2,167	2,126	1,200	1,347
Red River Entertainment of Shreveport, LLC	Sam's Town Hotel & Casino—Shreveport	22,649,968	864	854	674	517
St. Charles Gaming Company, Inc.	Isle of Capri Hotel Lake Charles—Westlake	24,023,916	967	938	491	559
Treasure Chest Casino, LLC	Treasure Chest Casino—Kenner	22,110,087	693	688	401	354
Bossier Casino Venture, Inc.	Margaritaville Resort Casino—Bossier City	30,977,477	1,032	1,021	710	641
Total		**$411,455,767**	**14,341**	**13,873**	**8,853**	**8,149**

D/B/A = doing business as.
[a]Average over calendar year.
[b]Golden Nugget Lake Charles, LLC opened December 2014.
Note: Dollar figure amounts are rounded to the nearest dollar.

SOURCE: "Riverboat Gaming Fiscal Year 2015–2016: Employment," in *21st Annual Report to the Louisiana State Legislature*, Louisiana Gaming Control Board, 2017, http://lgcb.dps.louisiana.gov/docs/2017_Annual_Report.pdf (accessed June 29, 2017)

populated and located in areas without other prospects for economic development, the effect of casinos on employment and wages is likely to resemble that of Tunica more closely than that of Atlantic City.

The growth of the casino employment market has spurred a related increase in vocational and professional training for casino workers. The University of Nevada, Las Vegas, which is only 1.5 miles (2.4 km) from the Strip, offers a major in gaming management that includes instruction in gaming operations, marketing, hospitality, security, and regulations. At Tulane University's School of Continuing Studies in New Orleans, students can choose from several programs that last between one and four years, including a bachelor's degree in casino resort management. Students pursuing a degree in hospitality and tourism management at the University of Massachusetts can specialize in casino management. Central Michigan University, which is located near the Soaring Eagle Casino and Resort operated by the Saginaw Chippewa Tribe, offers a business degree in gaming and entertainment management that includes coursework in the protection of casino table games, gaming regulations and control, the mathematics of casino games, and the sociology of gambling. The Casino Career Institute, which includes a large mock casino, is a division of Atlantic Cape Community College in downtown Atlantic City. When it opened in 1978, it was the first gaming school in the country to be affiliated with a community college.

Tribal Casinos and Federal Labor Laws

Historically, employees at tribal casinos have not been covered by the federal labor laws that protect workers at commercial casinos. As sovereign entities, tribes are expressly excluded from the definition of *employer* as specified by Title VII of the Civil Rights Act of 1964 and Title I of the Americans with Disabilities Act of 1990, which prohibit discrimination in employment on the basis of race, sex, physical impairment, and other criteria. Even in cases of clear-cut discrimination, tribes have successfully argued their immunity from criminal and civil penalties. For example, in "Indian Casinos Have Different Set of Laws" (SFExaminer.com, October 3, 2013), Christopher B. Dolan of the Dolan Law Firm of San Francisco, California, describes one such case, in which a client represented by his firm attempted to file suit against the Cache Creek Casino Resort owned and operated by the Yocha Dehe Wintun Nation. Dolan's client, an employee of the casino, was allegedly subjected to harassment based on race, sex, and national origin, and his attempts to report the behavior to casino management led to further harassment. The tribe pleaded sovereign immunity, and the case was thrown out of state court irrespective of the merits of the plaintiff's complaint.

Even so, tribal immunity from the National Labor Relations Act (NLRA), which established the right of employees to form unions and created the National Labor Relations Board (NLRB) to safeguard employee rights and combat unfair labor practices, has been curtailed since 2004. This change in precedent occurred as a result of the federal court case *San Manuel Indian Bingo and Casino* (341 NLRB 1055 [2004] aff'd. 475 F.3d 1306 [D.C. Cir. 2007]), which held that the San Manuel Indian Bingo and Casino in Southern California could not prevent the formation of a union by claiming sovereign immunity from the NLRA. *San Manuel Indian Bingo and Casino* was in turn used as a precedent to rule against the Mashantucket Pequot Tribe's attempt to prevent the United Auto Workers and the American Federation of Labor–Congress of Industrial Organizations from organizing dealers at the Foxwoods Casino and Resort in Connecticut. Although the tribe contested the legitimacy of the dealers' union, which was officially certified in June 2008, the NLRB repeatedly struck down the tribe's petitions, and the union remained intact as of 2017.

Kevin J. Allis of PilieroMazza PLLC in Washington, D.C., notes in "Federal Employment Laws Impact Tribal Employers" (2017, http://www.indiangaming.com/regulatory/view/?id=84) that federal courts in the 21st century typically consider tribal casinos subject to a number of other labor laws beyond the NLRA, including:

- The Fair Labor Standards Act, which establishes restrictions on employers regarding minimum wage, overtime pay, and child labor
- The Family and Medical Leave Act, which requires employers to provide eligible employees with up to 12 weeks of unpaid family and medical leave annually
- The Employee Retirement Income Security Act, which sets standards and provides protections for employee pension and health plans
- The Age Discrimination in Employment Act, which prohibits discrimination based on age against employees aged 40 years and older

TOURISM

In *2013 State of the States*, the most recent source of data on this topic as of August 2017, the AGA estimates that 71.6 million Americans, or nearly one-third (32%) of the 223.7 million Americans aged 21 years and older, gambled in a casino in 2012. More than three-quarters (79%) of adults who visited a casino in 2012 expected to visit a casino again at least once in the coming year, and 45% planned to make three or more trips to a casino. The casinos of Las Vegas and, to a lesser extent, Atlantic City attempt to provide a full-service resort experience that will attract tourist spending both on the casino floor and beyond it. Most other commercial and tribal casinos are designed with day-trippers from the surrounding area primarily in mind. These regional casinos do not generate substantial tourism revenues beyond the gaming floor, although some are located in major cities or areas that are tourist destinations for other reasons.

Las Vegas

No destination better represents the marriage between gambling and tourism than Las Vegas. According to the Las Vegas Convention and Visitors Authority (LVCVA), in "2016 Las Vegas Year-to-Date Executive Summary" (February 23, 2017, http://epubs.nsla.nv.gov/statepubs/epubs/651944-2016.pdf), 42.9 million tourists traveled to the city in 2016, and the citywide hotel occupancy rate for the entire year was 89.1%. A particularly important subset of visitors are those who travel to the city for corporate and professional conventions and trade shows. In 2016, 6.3 million people attended 21,864 conventions in the city.

In its bid to attract tourists, Las Vegas has had its ups and downs. During the early 1990s the city experienced a steep decline in revenues because of competition from legal gambling on riverboats and tribal casinos in other states. To counteract this development, the city began a drive to shift its focus from an adult playground to a family destination. As part of the campaign, $12 billion was spent to refurbish almost every hotel on the Strip and to add entertainment facilities. Theme hotels became a big draw. Adult entertainment along the Strip, such as topless shows, gave way to magic shows, circus events, and carnival rides. The LVCVA focused advertising on families. The result was a huge increase in visitors.

However, children distracted their parents from gambling. Casino owners noticed that the changes did not bring in more gambling revenue, so during the late 1990s the city began to change its image again. Adult entertainment made a comeback along the Strip: casino hotels began offering more topless and nude shows, although managers insisted that the nudity presented at their casinos would always be tasteful and artistic. They were careful not to offend shareholders of their parent corporations or to alienate women, who were potential gamblers and made up nearly 60% of Las Vegas visitors. In 2001 MGM Grand shut down its family theme park.

Theresa Howard reports in "Vegas Goes for Edgier Ads" (USAToday.com, August 3, 2003) that in early 2003 the LVCVA launched a somewhat risqué ad campaign with the slogan "What Happens Here, Stays Here." It may have been part of the reason that Las Vegas tourism and casino revenue increased substantially in 2005 and 2006. The focus on adults continued throughout the Great Recession, with MGM Grand launching the "Get Rewarded for Your Sins" promotion in October 2009. Amanda Finnegan explains in "Vegas Tourism Companies Embrace Social Media Strategies" (LasVegasSun.com, November 11, 2009) that the promotion encouraged visitors to recount their Vegas mishaps on Twitter and possibly win free hotel accommodations.

Regardless, the casino and tourism industries in Las Vegas were hit hard by the Great Recession. Convention attendance, which is a reliable source of casino spending each year, declined quite dramatically. In "Year End Summaries for 2004–2010" (May 20, 2014, http://www.lvcva.com/includes/content/images/media/docs/YE-Summaries-2004-2010.pdf), the LVCVA shows that the number of conventions held in the city fell from 23,847 in 2007 to 19,394 in 2009 and then bottomed out at 18,004 in 2010. The primary driver of this trend was reduced corporate spending on travel and other nonessential activities. However, Richard N. Velotta notes in "Signs of a Surge in Las Vegas Conventions" (LasVegasSun.com, December 28, 2010) that local tourism proponents also blamed President Barack Obama's (1961–) public declaration that companies taking federal stimulus funds (which were supplied by taxpayers) should not use those funds to travel to Las Vegas or the Super Bowl—the suggestion being that corporate travel to Las Vegas was inherently trivial or decadent. The convention numbers rebounded between 2010 and 2016. However, although the number of convention attendees returned to peak levels by 2016, the total number of conventions remained well below prerecession highs.

Atlantic City

Tourism in Atlantic City increased following the introduction of casino gambling, but not as fast or as much as many had hoped. From the 1880s to the 1940s Atlantic City was a major tourist destination, particularly for people living in the Northeast. Visitors went for the beaches and to walk along the town's boardwalk and piers, which featured carnival-like entertainment. However, during the 1960s the city fell into an economic depression as tourists ventured farther south to beaches in Florida and the Caribbean.

Casino gambling was legalized in 1976 in the hopes that the city would recapture its former glory and rival Las Vegas as a tourist destination. Progress was slow through the 1980s and early 1990s. Although visitors began going to Atlantic City, they mostly arrived by bus or car and stayed for only a day or two. In 1984 the state established the Casino Reinvestment Development Authority (CRDA) to revitalize the city using the funds from a 1.3% tax on casino revenues.

The economic troubles that had ravaged the city's businesses before gambling was legalized were not easily overcome. Vacant lots, dilapidated buildings, and housing projects surrounded the casinos. The overall atmosphere was not particularly appealing to vacationers or convention-goers. Mike Kelly notes in "Gambling with Our Future: City Poised to Hit Jackpot, or Lose Everything" (NorthJersey.com, July 1, 1993) that in 1993 the city was "trapped in a web of poverty and blight." At that time, the typical visitor was a retiree who arrived by bus and stayed for only the day. According to Kelly, Atlantic City's 30 million annual visitors actually represented about 5 million people making multiple trips.

During the late 1990s initiatives by the CRDA and other groups began paying off. Hundreds of new homes were built and commercial businesses were established. One of the largest convention centers in the country opened in May 1997. The city's image began to improve, and tourism showed a moderate surge.

However, the early 21st century proved extremely challenging for Atlantic City. The city's casinos experienced a decrease in attendance because of the recession, and smoking restrictions on the casino floor, which went into effect in 2008, also played a part. Additionally, the legalization of casino gambling in Pennsylvania and other nearby states cut sharply into Atlantic City's visitor totals between 2006 and 2017. Although the U.S. commercial casino industry as a whole contracted during the recession, Atlantic City's fortunes fell further and faster than those of the industry as a whole, and they showed little signs of rebounding. According to the New Jersey Casino Control Commission, in *Annual Report 2011* (June 2012, http://www.nj.gov/casinos/about/reports/pdf/2011_ccc_annual_report.pdf), casino gaming in Atlantic City generated $3.6 billion in 2010. As Table 4.4 in Chapter 4 shows, by 2016 Atlantic City gaming revenues fell to $2.4 billion.

CRIME

When gambling was legalized in Nevada in 1931, the law kept corporations out of the casino business by requiring that every shareholder obtain a gaming license. This law, which was designed to safeguard the integrity of the casinos, unintentionally gave organized crime a huge advantage. The nation was in the midst of the Great Depression (1929–1939), and building a flashy casino hotel was expensive. Few legitimate businesspeople had the cash to finance a casino, and banks were reluctant to lend money for something they considered to be a bad investment. Organized crime groups had made fortunes selling bootleg liquor during Prohibition, so they were able to make the capital investments needed to build and operate lavish casino hotels that attracted visitors.

The marriage between casinos and organized crime in Nevada lasted for decades but was eventually ended by gaming officials and law enforcement. In the 21st century there is little evidence of organized crime activity in the casino industry. Regulatory agencies keep a watchful eye on casinos to make sure mobsters and their associates do not gain a new foothold.

Casinos keep an equally watchful eye on their patrons and employees. For example, a host of security guards and cameras constantly monitors the casino floor. Observers watch dealers and patrons at the gaming tables and all money-counting areas. Some casinos use high-tech facial recognition programs to scan incoming patrons and quickly identify any known felons or other undesirables. Casinos are particularly alert to the possibility of illegal schemes for cheating at slot machines and table games, and state regulators keep public lists of those who are no longer allowed on casino grounds, typically because of their histories of cheating. For example, the Nevada Gaming Commission and the State Gaming Control Board maintain the "GCB Excluded Person List" (http://gaming.nv.gov/index.aspx?page=72), which listed 31 people as of May 2017. Although the industry does not release data on crimes committed by casino employees, analysts believe employee theft and embezzlement account for millions of dollars in losses each year. Vice crimes, particularly prostitution, and weapons crimes also occur.

According to Earl L. Grinols and David B. Mustard, in "Casinos, Crime, and Community Costs" (*Review of Economics and Statistics*, vol. 88, no. 1, February 2006), one of the most influential empirical studies on the relationship between casinos and crime, the amount of crime in a community with a casino has a direct relationship to the maturity of the casino. The researchers collected crime data from all 3,165 counties in the United States with and without casinos between 1977 and 1996. Their analysis shows that when a casino first opened in a county, crime changed very little, but slowly rose and then grew steadily in subsequent years. Although increased employment and expanded law enforcement reduced crime initially, over time these effects were overtaken by factors that were related to casinos. Grinols and Mustard explain, "Specifically, problem and pathological gamblers commit crimes as they deplete their resources, nonresidents who visit casinos may both commit and be victims of crime, and casino-induced changes in the population start small but grow." The researchers also note that "overall, 8.6% of property crime and 12.6% of violent crime [which includes robberies] in counties with casinos was due to the presence of the casino." Furthermore, they find "mixed evidence about whether casino openings increase neighbor-county crime rates."

Douglas M. Walker responds to Grinols and Mustard's study in "Do Casinos Really Cause Crime?" (*Econ Journal Watch*, vol. 5, no. 1, January 2008). He argues that Grinols and Mustard failed to address the effects of tourism on the crime rate, that the source of their data (the Uniform Crime Reports) is not appropriate for investigating longitudinal changes in local crime, and that they do not consider that casinos are sometimes introduced in high-crime areas. He concludes, "My point is not to suggest that casinos do not cause crime. They might.... However, the errors in the Grinols and Mustard study deserve attention because of the influence their study seems to be having among researchers, policy makers, the media, and voters."

SUICIDE

The possible link between casino gambling and suicide rates has been the subject of much investigation. For example, in December 1997 David P. Phillips, Ward R. Welty, and Marisa M. Smith concluded in "Elevated Suicide Levels Associated with Legalized Gambling" (*Suicide and Life-Threatening Behavior*, vol. 27, no. 4), an influential paper on the topic, that "visitors to and residents of major gaming communities experience significantly elevated suicide levels." The researchers found that in Atlantic City "abnormally high suicide levels" for visitors and residents appeared only after casinos opened.

However, five years later Richard McCleary et al. found in "Does Legalized Gambling Elevate the Risk of Suicide?" (*Suicide and Life-Threatening Behavior*, vol. 32, no. 2, Summer 2002) little to no correlation between suicide rates and the presence of casino gambling in U.S. communities. After comparing the 1990 suicide rates of 148 metropolitan areas in different regions of the country, the researchers found that the presence of casinos could account for only 1% of the regional differences in suicide rates. They also compared "before and after" suicide rates for cities in which gambling had been legalized. Although increased suicide rates were noted in Atlantic County, New Jersey, and Harrison County, Mississippi, after the advent of gambling, the increases were not considered to be statistically significant. McCleary et al. noted that suicide rates dropped significantly in Lawrence County, South Dakota, after casino gambling was introduced in the town of Deadwood.

In "Risk Factors for Suicide Ideation and Attempts among Pathological Gamblers" (*American Journal on Addictions*, vol. 15, no. 4, July–August 2006), David C. Hodgins, Chrystal Mansley, and Kylie Thygesen find that suicidal ideation and suicide attempts are more likely among pathological gamblers. However, the history of suicidal thoughts generally preceded problem gambling behavior by an average of more than 10 years. The researchers conclude that previous mental health disorders, such as clinical depression, put individuals more at risk for both suicide and gambling problems. Paul W. C. Wong et al. find in "A Psychological Autopsy Study of Pathological Gamblers Who Died by Suicide" (*Journal of Affective Disorders*, vol. 120, no. 1, January 2010) that a high proportion of suicide cases in which pathological gambling was present had a comorbid (an additional or coexisting disease) psychiatric illness, most often depression, at the time of death. In "The Clustering of Psychiatric Disorders in High-Risk Gambling Populations" (*Journal of Gambling Studies*, June 18, 2013), Mohammad Reza Abdollahnejad, Paul Delfabbro, and Linley Denson similarly find, based on surveys of regular gamblers, that nearly two-thirds of pathological gamblers reported "both an anxiety or mood disorder in conjunction with another type of disorder." Accordingly, the researchers suggest that the overlap between those disorders and disordered gambling predisposes such individuals to consider suicide when faced with negative gambling outcomes. Additionally, the link between suicidal ideation and problem gambling is not limited to gamblers themselves. Donald W. Black et al. of the University of Iowa Carver College of Medicine reveal in "Suicide Ideations, Suicide Attempts, and Completed Suicide in Persons with Pathological Gambling and Their First-Degree Relatives" (*Suicide & Life-Threatening Behavior*, vol. 45, no. 6, December 2015) that children of pathological gamblers have significantly higher rates of suicide attempts than children of nongamblers.

BANKRUPTCY

Establishing a definitive link between gambling habits and bankruptcy is difficult. As with criminal behavior and suicide, it is likely that disordered gamblers are at a significantly higher risk of bankruptcy than casual gamblers. However, attempts to show a correlation in the aggregate between the prevalence of casino gambling and bankruptcy have yielded inconclusive results over the years.

In "Estimating the Effects of Casinos and of Lotteries on Bankruptcy: A Panel Data Set Approach" (*Journal of Gambling Studies*, vol. 27, no. 1, March 2011), Bogdan Daraban and Clifford F. Thies note that eight prominent academic studies published between 1999 and 2006 address the question of whether the legalization of gambling in a given locality led to an accompanying increase in bankruptcy filings, and only two found a substantial correlation. Daraban and Thies conduct their own investigation into the question by considering the correlation between state revenue generated from gambling and bankruptcy filings. The researchers conclude that the presence of casino gambling increases bankruptcy filings by 2%, which is slightly higher than previous estimates that found a correlation between the two factors.

Jon E. Grant et al. of the University of Minnesota School of Medicine study in "Pathologic Gambling and Bankruptcy" (*Comprehensive Psychiatry*, vol. 51, no. 2, March–April 2010) 517 pathological gamblers and find that there were specific clinical differences between those who declared bankruptcy and those who did not. The researchers report that gamblers who declared bankruptcy were more likely to be single and to have work, marital, or legal problems secondary to gambling. These gamblers also tended to become problem gamblers at an earlier age and were more likely to have comorbid depressive disorders or substance use disorders than were gamblers who did not declare bankruptcy.

DISORDERED GAMBLING

Although disordered gambling, also known as pathological or compulsive gambling, is a real problem with significant social costs, the overwhelming majority of gamblers play for recreation and practice healthy gambling behaviors. In *2013 State of the States*, the AGA notes that in 2012, 86% of casino gamblers set a budget before they began playing and that 53% of casino gamblers set a daily budget of $100 or less. According to the National Center for Responsible Gaming, in *Increasing the Odds: Volume 3, Gambling and the Public Health, Part 1* (2009, http://www.ncrg.org/sites/default/files/uploads/docs/faq/ncrg_monograph_vol3.pdf), a survey of dozens of studies dating from the 1970s to the early 21st century and that examines the prevalence rates of pathological gambling (the proportion of the adult population that has exhibited pathological gambling behavior in the past), researchers have consistently estimated the prevalence of pathological gambling at between 0.2% and 2.1%. The center receives most of its funding from the commercial casino industry, but the consistency of these estimates worldwide, over time, and across various study methodologies has been widely noted.

Self-Exclusion Programs

Many casinos operate self-exclusion programs in which people can voluntarily ban themselves from casinos. A number of states also offer self-exclusion programs for all casinos within their borders. For example, Missouri's Voluntary Exclusion Program (http://www.mgc.dps.mo.gov/DAP/dap_onList.html) was created in 1996 after a citizen requested that he be banned from the riverboat casinos because he was unable to control his gambling. The Missouri Gaming Commission requires that the casinos remove self-excluded people from their direct marketing lists, deny them check-cashing privileges and membership in players' clubs, and cross-check for their names on the self-exclusion list before paying out any jackpots of $1,200 or more. The casinos are not responsible for barring listed people, but anyone listed is to be arrested for trespassing if he or she violates the ban and is discovered in a casino. However, self-excluded people can enter the casino for employment purposes.

Sara E. Nelson et al. study the effectiveness of Missouri's Voluntary Exclusion Program in "One Decade of Self Exclusion: Missouri Casino Self-Excluders Four to Ten Years after Enrollment" (*Journal of Gambling Studies*, vol. 26, no. 1, March 2010). The researchers find that most gamblers who enrolled in the self-exclusion program had been able to reduce their gambling. However, half of the self-excluders who had attempted to enter Missouri casinos after they enrolled in the program were still able to do so. Nelson et al. conclude that the benefits of the program are more attributable to the act of enrollment than in the casinos' actual enforcement of the ban. In addition, the researchers suggest that self-excluders who sought additional mental health treatment were better able to control their gambling problem and that self-exclusion programs should provide information about additional help and treatment options at the time of enrollment.

Programs in other states are similar. If a self-excluded person is discovered in an Illinois casino, his or her chips and tokens are taken away and their value is donated to charity. The Illinois self-exclusion program runs for a minimum of five years. After that time, people can be removed from the program if they provide written documentation from a licensed mental health professional that they are no longer problem gamblers. Self-exclusion in Michigan is permanent; a person who chooses to be on the Disassociated Persons List is banned for life from Detroit casinos. In New Jersey the Casino Control Commission allows people to voluntarily suspend their credit privileges at all Atlantic City casinos. The commission maintains a list of those who have joined the program and shares the list with the casinos.

Besides casinos and states, companies that provide the automatic teller machines and cash-advance services for casinos have put self-exclusion programs into place. For example, Global Payments provides self-exclusion and self-limit services for people with gambling problems. Those who put their names on the self-exclusion list are denied money or cash advances, whereas the self-limit program restricts how much money patrons can withdraw during a specified period.

Hotlines and Treatment

All states operate gambling hotlines that either refer callers to other groups for help or provide counseling over the phone. Gamblers Anonymous provides a list of hotlines by state on its website U.S. Hotlines (http://www.gamblersanonymous.org/ga/hotlines). The National Council on Problem Gambling operates the National Problem Gambling Helpline Network (http://www.ncpgambling.org/help-treatment/national-helpline-1-800-522-4700), which consists of 28 call centers that direct callers to resources in all 50 states.

Most states have nonprofit or government agencies devoted to the prevention and treatment of problem gambling, and many release public information about the use of their hotlines and treatment programs. For example, the Indiana Council on Problem Gambling (ICPG) tracks the volume of calls to the state-funded Indiana hotline and the number of problem gamblers who subsequently enter treatment. According to the ICPG, in "State Funded Help Line Calls and State Funded Problem Gambling Treatment Enrollment Data by Indiana County by Year" (April 27, 2017, http://www.indianaproblemgambling.org/PDF/county_data_thru_FY16.pdf), the state-funded hotline received 320 calls during FY 2016.

The Maryland Center of Excellence on Problem Gambling was founded by the University of Maryland School of Medicine to help provide support for individuals struggling with disordered gambling. Besides maintaining a 24-hour helpline, the center offers training programs for counselors and public health professionals, sponsors outreach initiatives, conducts research, and develops treatment programs. In *Expanding the Vision: FY 2016 Annual Report* (2016, https://bha.health.maryland.gov/Documents/FY16CPGAnnual%20Report.FINAL. pdf), the center provides data relating to its counseling and educational activities. During FY 2016 (July 1, 2015, to June 30, 2016) the center reports that it received 711 calls, texts, or online chat messages from individuals seeking help for disordered gambling. (See Figure 6.3.) This figure represented a 65% increase over the number of communications received during FY 2013, when the center received 431 calls, texts, and chat messages. As Figure 6.4 shows, the majority of the people who contacted the center during FY 2016 reported that their primary gambling problem involved either casino table games or casino slot machines.

According to Figure 6.5, more women than men contacted the Maryland Center of Excellence on Problem Gambling during FY 2013. Male callers outnumbered female callers by substantial margins during FYs 2014 and 2015, and by a smaller margin during FY 2016. Roughly equal numbers of whites and African Americans contacted the center during FY 2016. (See Figure 6.6.) Taken together, these groups accounted for approximately

FIGURE 6.3

Annual number of phone, text, and online chat contacts received by Maryland Problem Gambling Helpline, fiscal years 2013–16

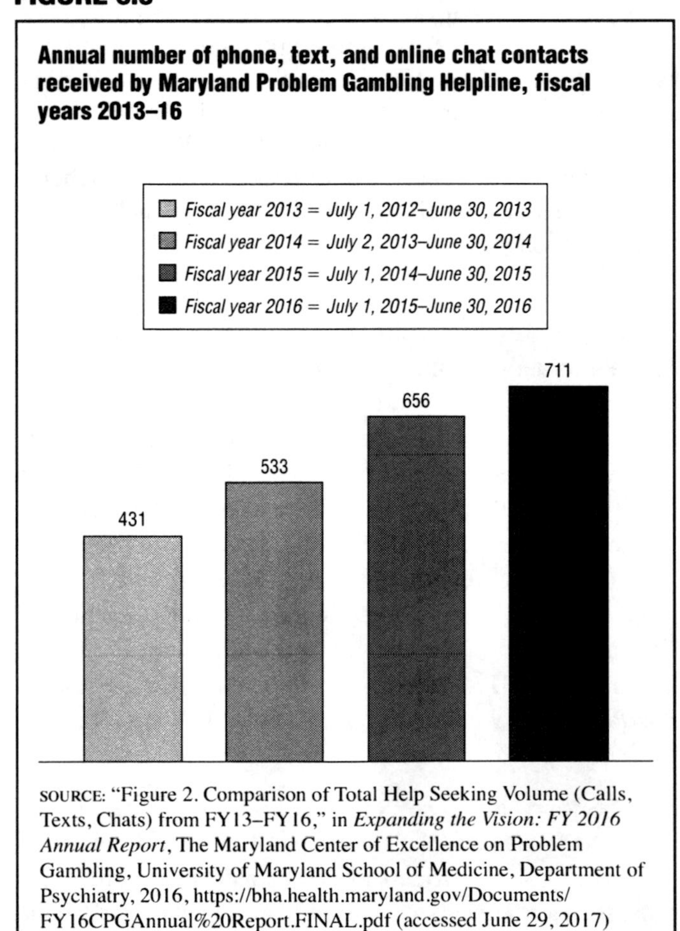

SOURCE: "Figure 2. Comparison of Total Help Seeking Volume (Calls, Texts, Chats) from FY13–FY16," in *Expanding the Vision: FY 2016 Annual Report*, The Maryland Center of Excellence on Problem Gambling, University of Maryland School of Medicine, Department of Psychiatry, 2016, https://bha.health.maryland.gov/Documents/FY16CPGAnnual%20Report.FINAL.pdf (accessed June 29, 2017)

500 of the 711 calls received that year. The center reports that nearly three-quarters (74%) of calls received at the helpline came from the gamblers themselves; the remaining quarter of communications came from parents, children, spouses, or other individuals. (See Figure 6.7.)

UNDERAGE GAMBLING

The legal gambling age in all commercial casinos in the United States is 21 years; in tribal casinos it varies from 18 to 21 years.

Unlike most other gaming states, Nevada allows escorted children to walk through casinos. Nevertheless, the state's casinos are prohibited from attempting to promote gambling activity among minors. The Nevada Gaming Commission's Regulation 14 (March 2017, http://gaming.nv.gov/modules/showdocument.aspx?documentid=2921) bans slot machines "derived from or based on a product that is currently and primarily intended or marketed for use by persons under 21 years of age." In "AGA Code of Conduct for Responsible Gaming" (2017, https://www.americangaming.org/sites/default/files/AGA%20Code%20of%20Conduct%20for%20Responsible%20Gaming_Final%207.27.17.pdf), the AGA similarly provides guidelines for ensuring that minors do not gamble in casinos. For example, it notes that casinos are expected to stop any minor from loitering on the casino floor, and casino employees are to be trained to deal with minors who attempt to buy alcohol or gamble.

Casinos appear to be successful in following the guidelines, and casino gambling among underage youth is not widely considered a significant problem. For example, in *2016 Michigan Gaming Control Board Annual Report* (April 2017, http://www.michigan.gov/documents/mgcb/2016_MGCB_Annual_Report_Public_Version_FINAL_558221_7.pdf), the Michigan Gaming Control Board provides statistics on minors attempting to enter Detroit's three casinos. A total of 24,957 minors tried to enter MGM Grand (2,209), MotorCity Casino (10,833), and Greektown Casino (11,915) casinos and were denied entry in 2016. (See Table 6.4.) Only a handful of minors were detected using slot machines, playing table games, or consuming alcohol in the three casinos; and only 31 were taken into custody by law enforcement officers. By contrast, the Missouri Gaming Commission reveals in *Annual Report 2016* (September 2016, http://www.mgc.dps.mo.gov/annual_reports/AR_CurrentYear/00_FullReport.pdf) that the Missouri highway patrol made a total of 5,499 arrests at state casinos during FY 2016. (See Table 6.5.) Although the latter report does not provide data concerning arrests of minors, a comparison of statistics from Michigan and Missouri seems to indicate that minors constitute an extremely small proportion of casino arrests in the United States. These findings suggest that much of the prevention of underage gambling occurs during attempted entry, when most minors are turned away from casinos.

FIGURE 6.4

Gambling problems reported by callers to Maryland Problem Gambling Helpline, fiscal years 2013–16

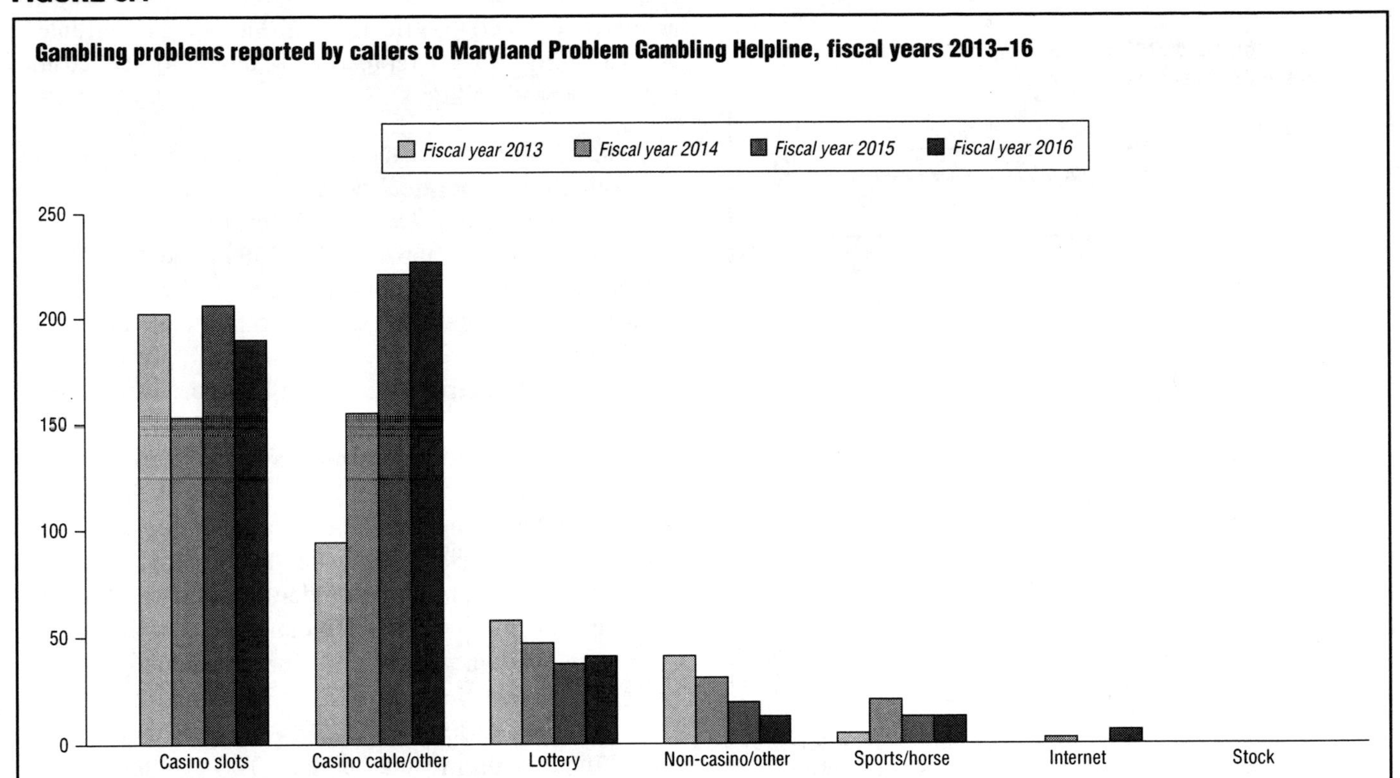

SOURCE: "Figure 6. Primary Gambling Problem Reported by Helpline Callers FY13–FY16," in *Expanding the Vision: FY 2016 Annual Report*, The Maryland Center of Excellence on Problem Gambling, University of Maryland School of Medicine, Department of Psychiatry, 2016, https://bha.health.maryland.gov/Documents/FY16CPGAnnual%20Report.FINAL.pdf (accessed June 29, 2017)

FIGURE 6.5

Callers to Maryland Problem Gambling Helpline, by gender, fiscal years 2013–16

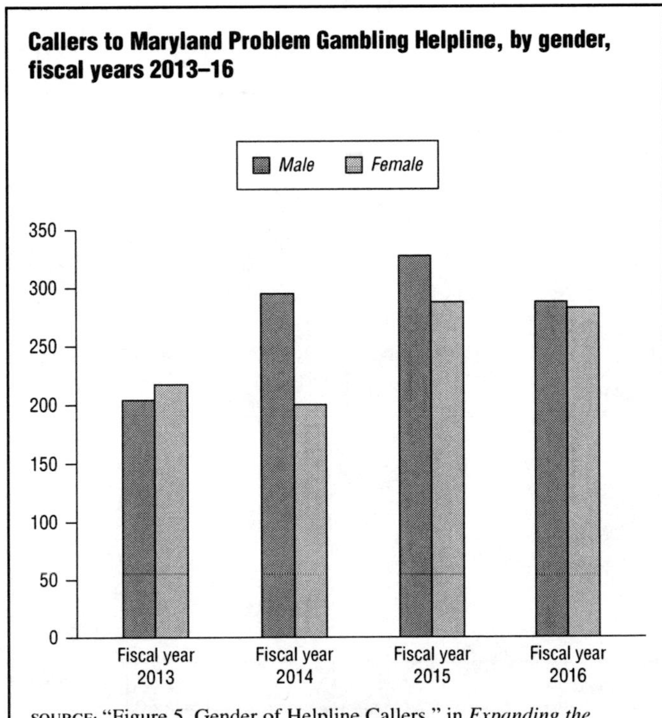

SOURCE: "Figure 5. Gender of Helpline Callers," in *Expanding the Vision: FY 2016 Annual Report*, The Maryland Center of Excellence on Problem Gambling, University of Maryland School of Medicine, Department of Psychiatry, 2016, https://bha.health.maryland.gov/Documents/FY16CPGAnnual%20Report.FINAL.pdf (accessed June 29, 2017)

POLITICS

Gambling and politics have always been linked, largely because casinos and other gaming establishments are so heavily regulated, the number of licenses available is often limited, and a great deal of money can be made by the people who hold those licenses. Lobbying (a common factor in the political system) can easily turn into influence peddling and bribery at all levels of government.

Douglas M. Walker and Peter T. Calcagno of the College of Charleston establish in "Casinos and Political Corruption in the United States: A Granger Causality Analysis" (*Applied Economics*, vol. 45, no. 34, 2013) the first empirical evidence that casinos are linked to political corruption. Their research and economic modeling suggests that, based on data from 1985 to 2000, those who are charged with overseeing the casino industry are subject to "regulatory capture" (a situation in which an agency created to regulate an industry for the good of the public has begun acting in ways that benefit the industry rather than the public). The researchers note that their findings may explain the tendency of gaming regulators to soften regulations on the industry as time passes beyond the initial legalization of casino gambling.

Both commercial and tribal casino operators make substantial campaign donations and employ lobbying firms to influence policy, typically with a goal of minimizing

FIGURE 6.6

Callers to Maryland Problem Gambling Helpline, by race/ethnicity, fiscal years 2013–16

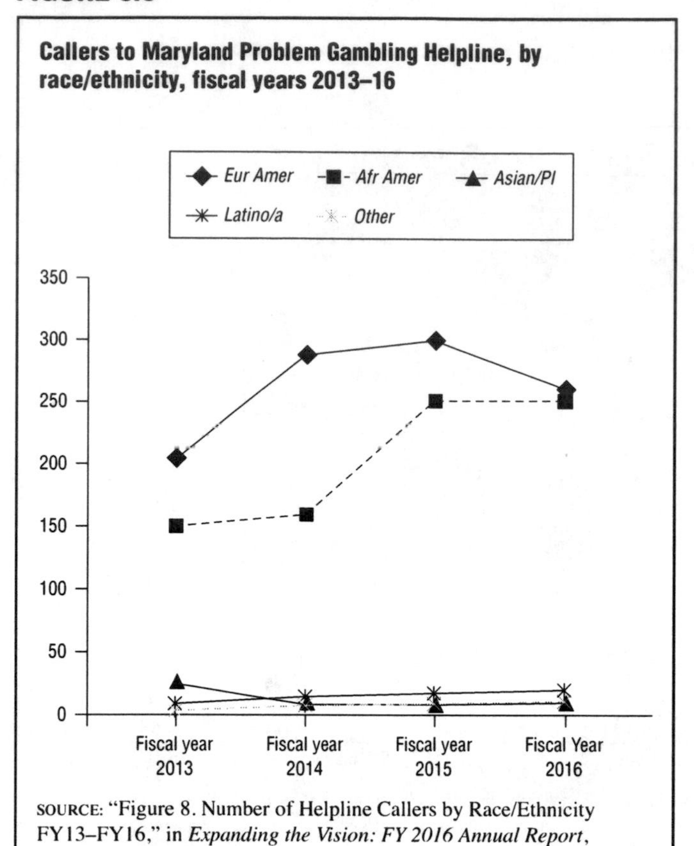

SOURCE: "Figure 8. Number of Helpline Callers by Race/Ethnicity FY13–FY16," in *Expanding the Vision: FY 2016 Annual Report*, The Maryland Center of Excellence on Problem Gambling, University of Maryland School of Medicine, Department of Psychiatry, 2016, https://bha.health.maryland.gov/Documents/FY16CPGAnnual%20Report.FINAL.pdf(accessed June 29, 2017)

casino regulation. Commercial and tribal casinos have also actively attempted to influence legislation regarding Internet gambling, which represents a threat to the casino industry's profitability.

The Center for Responsive Politics, a nonpartisan group that tracks political donations, indicates in "Casinos/Gambling" (May 16, 2017, https://www.opensecrets.org/industries/indus.php?ind=N07) that the Las Vegas Sands Corporation, a publicly traded casino and resort company, contributed $44 million to political campaigns in 2015–16. This sum dwarfed the total given by the second-largest contributor, Station Casinos, which spent $3.1 million on political donations that year. The Center for Responsive Politics notes that Native American tribes are among the top casino-industry donors at the federal level. The Oneida Indian Nation made the most political donations of any tribal casino–affiliated entity in 2015–16, contributing $2 million to federal candidates. Furthermore, the Puyallup Tribe of Indians ($1.1 million), the Chickasaw Nation ($819,574), the Pechanga Band of Luiseno Mission Indians ($734,631), and the Poarch Band of Creek Indians ($733,790) all contributed more than $700,000 during that period. The politicians who received the most gambling donations in 2015–16 were the Democratic presidential candidate Hillary Rodham Clinton (1947–); the U.S. representatives Tom Cole (1949–; R-OK), Joe Heck (1961–; R-NV), and Ben R. Luján (1972–; D-NM); and the U.S. senator Catherine Cortez Masto (1964–; D-NV).

FIGURE 6.7

Callers to Maryland Problem Gambling Helpline, by identity of caller, fiscal year 2016

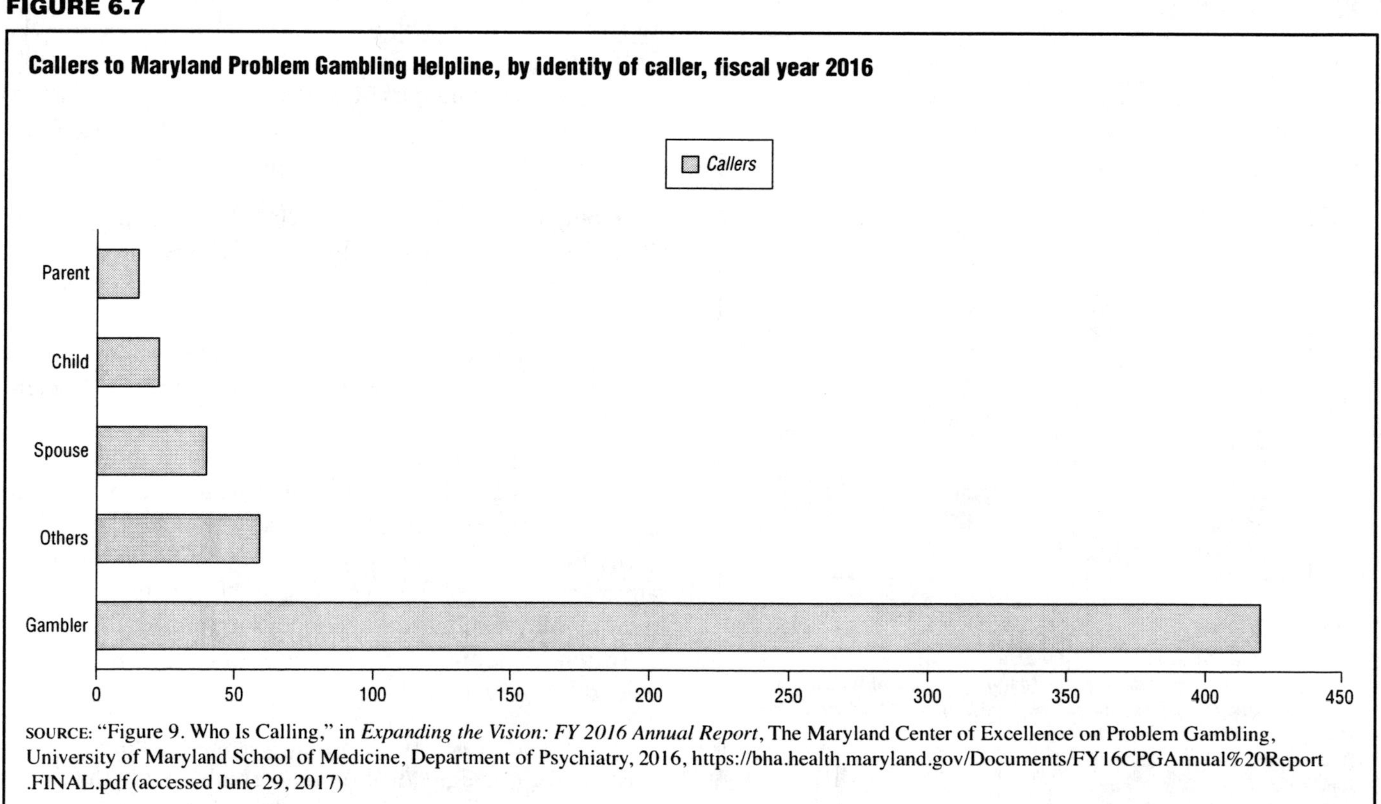

SOURCE: "Figure 9. Who Is Calling," in *Expanding the Vision: FY 2016 Annual Report*, The Maryland Center of Excellence on Problem Gambling, University of Maryland School of Medicine, Department of Psychiatry, 2016, https://bha.health.maryland.gov/Documents/FY16CPGAnnual%20Report.FINAL.pdf (accessed June 29, 2017)

Besides political donations, casino operators spend large sums of money employing lobbyists who also attempt to influence the political and legislative processes. According to the Center for Responsive Politics, during the first half of 2017 the Gila River Indian Community spent $1.3 million on lobbying, followed by the AGA ($780,000), the Caesars Entertainment Corporation ($739,000), MGM Resorts International ($700,000), and the Tohono O'odham Nation ($540,000). Between 2012 and 2016 the industry spent $174.2 million on lobbying at the federal level.

Some jurisdictions have become so concerned about the confluence of political pressure and money that they prohibit casino license applicants from making contributions to political candidates. Mississippi decided to prevent this temptation by not setting a limit on the number of casinos that can be built. State officials claimed their policy would prevent the bribery, extortion, and favoritism that had plagued neighboring Louisiana, where the number of licenses available for riverboat casinos was set at 15. Those licenses were so highly prized that Governor Edwin Edwards (1927–) extorted $3 million from people who wanted them. In May 2000 he was convicted of racketeering, extortion, and fraud and sentenced to 10 years in prison.

The intersection of politics and gambling at the federal level resulted in one of the most high-profile political corruption cases in recent decades, involving Jack Abramoff (1958–), a successful Washington, D.C., lobbyist. As a result of his involvement in tribal and commercial gaming, Abramoff pleaded guilty to fraud, tax evasion, and conspiracy to bribe public officials in January 2006. He was sentenced to 70 months in prison and ordered to pay $21 million in restitution.

Many credit Abramoff and his colleagues with securing the defeat of the Internet Gambling Prohibition Act of 1999. The bill was one of the first anti–Internet gambling bills proposed in Congress. It was passed in the U.S. Senate in 1999 and was put forth in the U.S. House of Representatives the following year. At the time, Abramoff was working for eLottery, an Internet site that wanted to sell state lottery tickets online. Its business was threatened by the legislation, so Abramoff gave money to conservative special interest groups to get them to pressure conservative House members to drop the bill because it contained exceptions for horse racing and jai alai. Through procedural maneuvering, a two-thirds majority was needed to pass the bill; it failed. When the bill's original supporters demanded that it be revived, Abramoff targeted 10 Republican House members in vulnerable districts with media and direct-mail campaigns that accused them of being "soft on gambling" if they voted for the bill. The representatives received so much pressure from their constituents that the House

TABLE 6.4

Contacts with minors in Detroit, Michigan, casinos, 2016

Number of minors	MGM Grand	MotorCity	Greektown
Denied entry into the casino	2,209	10,833	11,915
Physically escorted from the casino premises	27	3	3
Detected participating in gambling games other than slot machines	2	0	0
Detected using slot machines	3	1	0
Taken into custody by a law enforcement agency on the casino premises	27	4	0
Detected illegally consuming alcohol on the casino premises	5	0	0

SOURCE: "Casino Licensees' Reported Contacts with Minors on Licensed Casino Premises during Calendar Year 2016," in *2016 Michigan Gaming Control Board Annual Report*, Michigan Gaming Control Board, 2017, http://www.michigan.gov/documents/mgcb/2016_MGCB_Annual_Report_Public_Version_FINAL_558221_7.pdf (accessed June 29, 2017)

TABLE 6.5

Arrests at Missouri casinos, by month, fiscal year 2016

Boat	Jul	Aug	Sep	Oct	Nov	Dec	Jan	Feb	Mar	Apr	May	Jun	Totals
Ameristar KC	97	82	40	40	49	84	98	96	79	81	51	59	856
Ameristar SC	40	22	31	17	42	13	14	28	22	25	13	22	289
Argosy	37	79	23	25	34	38	26	25	60	27	31	31	436
Harrah's NKC	22	21	36	40	32	49	22	48	10	31	15	24	350
Hollywood—STL	25	50	16	18	10	19	19	15	12	12	40	20	256
Isle of Capri—Boonville	6	7	4	40	1	7	2	3	9	6	3	1	89
Isle of Capri—Cape	5	10	5	10	14	7	10	20	13	21	14	4	133
Isle of Capri—KC	23	35	31	27	22	37	45	15	59	27	42	30	393
Lady Luck/IOC	1	3	0	1	3	0	0	0	0	0	2	0	10
Lumiere Place	150	80	94	137	95	155	228	226	177	170	194	178	1,884
Mark Twain	3	2	2	6	2	2	0	1	8	2	2	5	35
River City	92	64	62	56	40	38	45	52	49	67	85	79	729
St. Jo Frontier	5	2	2	2	2	7	4	7	1	5	1	1	39
Totals	**506**	**457**	**346**	**419**	**346**	**456**	**513**	**536**	**499**	**474**	**493**	**454**	**5,499**

SOURCE: "2016 MSHP-GD Fiscal Year Casino Arrest," in *Annual Report 2016*, Missouri Gaming Commission, 2017, http://www.mgc.dps.mo.gov/annual_reports/AR_CurrentYear/00_FullReport.pdf (accessed June 29, 2017)

Republican leadership, fearing that the party might lose four seats in the 2000 election, decided not to bring the bill up for another vote.

Later in his career Abramoff and his team defrauded Native American tribes out of millions of dollars. Typically, he promised that, as their lobbyist, he could secure funding from the government for special projects, such as wider roads or new schools, and that he could keep the government from interfering in their operations, including casinos. In return, the tribes paid his lobbying firm and a public relations company more than $85 million.

In some instances Abramoff worked against a tribe behind the scenes and then offered to help it out for huge sums of money. For example, in 2002 he and his colleagues were instrumental in shutting down the Speaking Rock Casino in El Paso, Texas. He then went to the Tigua Tribe, which operated the casino, and claimed that he and his colleagues could convince Congress to reopen the casino. The tribe paid $4.2 million in lobbying fees, but the casino remained closed. After years of legal battling, the facility reopened in 2010 as the Speaking Rock Entertainment Center, offering free concerts to attract patrons to its restaurant, bar, and limited video gaming machines.

CHAPTER 7
LOTTERIES

A lottery is a game of chance in which people pay for the opportunity to win prizes. Part of the money taken in by a lottery is used to award the winners and to pay the costs of administering the lottery. The money left over is profit. Lotteries are extremely popular and legal in more than 100 countries.

In the United States all lotteries are operated by state governments, which have granted themselves the sole right to do so. In other words, they are monopolies that do not allow any commercial lotteries to compete against them. The profits from U.S. lotteries are used solely to fund government programs. As of 2017, lotteries operated in 44 states and the District of Columbia. In addition, lottery tickets could be legally purchased by any adult physically present in a lottery state, even if that adult did not reside in the state.

According to the U.S. Census Bureau, state lotteries took in $66.8 billion in revenue and paid out $42.3 billion in prizes in 2015. (See Table 7.1.) After paying winners, state lotteries spent $3.2 billion on administration costs, and the remaining $21.4 billion represented state government profits. Although high-population states in general generate more lottery sales and profits than low-population states, the relationship between population and sales is not exact. New York generated $7.9 billion in lottery sales, far more than any other state. California ranked second in terms of lottery sales, with $5.5 billion, followed by Florida ($5.3 billion), Massachusetts ($5 billion), and Texas ($4.3 billion). One factor in the disproportionate size of sales in New York (the third-largest state by population) and Massachusetts (the 14th-largest state by population) may be that both states were among the earliest in the country to establish lotteries. The New York Lottery began in 1967, and Massachusetts instituted a lottery in 1972. In general, western and southern states adopted lotteries later than eastern and midwestern states. As of August 2017, the six states that did not have lotteries were Alabama, Alaska, Hawaii, Mississippi, Nevada, and Utah. The Wyoming Lottery, which began in 2013, was the newest state to operate a lottery.

LOTTERY HISTORY

Early History

The drawing of lots to determine ownership or other rights is recorded in many ancient documents. The practice became common in Europe during the late 15th and early 16th centuries. Lotteries were first tied directly to the United States in 1612, when King James I (1566–1625) of England created a lottery to provide funds to Jamestown, Virginia, the first permanent British settlement in North America. Lotteries were used by public and private organizations after that time to raise money for towns, wars, colleges, and public works projects.

During the 1760s George Washington (1732–1799) conducted a lottery to finance the construction of the Mountain Road in Virginia. Benjamin Franklin (1706–1790) supported lotteries to pay for cannons during the American Revolution (1775–1783). John Hancock (1737–1793) ran a lottery to finance the rebuilding of Faneuil Hall in Boston, Massachusetts, after it was nearly destroyed by fire in 1761. Lotteries fell into disfavor during the 1820s because of concerns that they were harmful to the public. New York was the first state to pass a constitutional prohibition against them.

The Rise and Fall of Lotteries

The southern states relied on lotteries after the Civil War (1861–1865) to finance Reconstruction (1865–1877). The Louisiana lottery, in particular, became widely popular. According to the Louisiana Lottery Corporation, in "History of Lotteries" (2017, http://www.louisianalottery.com/history-of-lotteries), in 1868 the Louisiana Lottery Company was granted permission by the state legislature to operate tax free provided that it made a donation

TABLE 7.1

Lottery income and apportionment of funds, by state, 2015

[Dollars in thousands]

State	Income (ticket sales excluding commissions)	Apportionment of funds		
		Prizes	Administration	Proceeds available
United States	66,788,035	42,278,889	3,180,173	21,362,759
Alabama	—	—	—	—
Alaska	—	—	—	—
Arizona	698,969	486,638	41,994	170,337
Arkansas	385,409	280,466	32,989	71,954
California	5,524,851	3,501,746	270,567	1,752,538
Colorado	498,210	331,499	40,043	126,668
Connecticut	1,079,703	707,735	44,060	327,908
Delaware	400,647	108,964	50,942	240,741
Florida	5,277,739	3,627,939	156,890	1,492,910
Georgia	3,674,014	2,528,871	164,678	980,465
Hawaii	—	—	—	—
Idaho	193,831	136,769	11,118	45,944
Illinois	2,837,805	1,817,478	308,686	711,641
Indiana	970,535	670,980	143,348	156,207
Iowa	324,767	196,882	51,525	76,360
Kansas	231,272	138,741	23,429	69,102
Kentucky	831,073	556,276	41,419	233,378
Louisiana	427,181	239,198	25,769	162,214
Maine	236,179	166,667	14,165	89,133
Maryland	2,131,494	1,051,486	85,581	994,427
Massachusetts	5,005,635	3,641,351	100,590	1,263,694
Michigan	2,525,711	1,696,978	73,135	755,598
Minnesota	514,022	351,668	25,036	137,318
Mississippi	—	—	—	—
Missouri	1,107,013	755,429	46,746	304,838
Montana	55,451	29,256	8,227	17,968
Nebraska	149,743	94,696	18,357	36,690
Nevada	—	—	—	—
New Hampshire	281,133	176,415	16,005	88,713
New Jersey	2,831,083	1,873,303	119,337	838,443
New Mexico	137,017	75,592	20,418	41,007
New York	7,873,768	4,396,853	374,262	3,102,653
North Carolina	1,834,453	1,231,233	81,448	521,772
North Dakota	25,841	13,978	5,011	6,852
Ohio	2,712,802	1,875,258	122,226	715,318
Oklahoma	171,634	87,783	24,053	59,798
Oregon	901,728	211,444	77,975	612,309
Pennsylvania	3,546,607	2,411,651	39,998	1,094,958
Rhode Island	542,736	150,063	10,076	382,597
South Carolina	1,302,825	924,137	38,827	339,861
South Dakota	147,933	29,744	7,753	110,436
Tennessee	1,378,987	774,250	59,145	545,592
Texas	4,281,136	2,858,319	191,129	1,231,688
Utah	—	—	—	—
Vermont	104,861	72,710	9,228	22,923
Virginia	1,739,959	1,116,625	87,638	535,696
Washington	658,854	432,901	45,142	180,811
West Virginia	658,793	106,476	32,308	520,009
Wisconsin	574,631	342,441	38,900	193,290
Wyoming	—	—	—	—

Note: A dash (—) indicates that the state government did not administer a lottery program.
Created: May 11, 2017.
Last Revised: May 11, 2017.

SOURCE: "Income and Apportionment of State-Administered Lottery Funds: 2015, in *2015 Annual Survey of State Government Finances*, U.S. Census Bureau, May 11, 2017, https://www2.census.gov/govs/state/2015_lottery_table.xls (accessed June 29, 2017)

of $40,000 per year for 25 years to the Charity Hospital of New Orleans. The company sold tickets nationwide and was one of the largest businesses in the United States. Charges of corruption eventually forced it to close. In 1890 Congress banned the mailing of lottery materials, and five years later it passed a law against the transport of lottery tickets across state lines. Following the Louisiana Lottery Company's closure the public learned that the lottery had been operated by a northern crime syndicate that regularly bribed legislators and committed widespread deception and fraud. The resulting scandal contributed to the turning of public opinion against lotteries. By 1894 no lotteries were legally operating in the United States.

Negative attitudes about gambling began to soften during the early 20th century, particularly after the failure of Prohibition (1920–1933). The state of Nevada legalized casino gambling during the early 1930s, and gambling for charitable purposes slowly became more commonplace across the country. Still, lingering fears about fraud kept public sentiment against lotteries for two more decades.

Rebirth during the 1960s

In 1963 the New Hampshire legislature authorized a sweepstakes to raise revenue for education, and in 1964 the New Hampshire Lottery began operating, becoming the first modern state-run lottery. Patterned after the popular Irish Sweepstakes, the first game offered in New Hampshire was much different from the lotteries of the 21st century. Tickets were sold for $3, and drawings were held infrequently. The biggest prizes were tied to the outcomes of particular horse races at the Rockingham Park racetrack in Salem, New Hampshire. Nearly $5.7 million was wagered during the lottery's first year.

In 1967 New York became the second state to institute a lottery, and the effort was an immediate success, grossing $53.6 million during its first year. The New York Lottery also enticed residents from neighboring states to cross state lines and buy tickets. Twelve other states established lotteries during the 1970s (Connecticut, Delaware, Illinois, Maine, Maryland, Massachusetts, Michigan, New Jersey, Ohio, Pennsylvania, Rhode Island, and Vermont). Analysts suggest that lotteries became firmly entrenched throughout the Northeast for three reasons: each state needed to raise money for public projects without increasing taxes, each state had a large Catholic population that was generally tolerant of gambling, and history shows that states are more likely to start a lottery if one is already offered in a nearby state. For example, Ron Stodghill and Ron Nixon report in "For Schools, Lottery Payoffs Fall Short of Promises" (NYTimes.com, October 7, 2007) that Mike Easley (1950–), the governor of North Carolina, said before his state established a lottery, "Our people are playing the lottery. We just need to decide which schools we should fund, other states' or ours."

During the 1980s lottery fever spread south and west. Seventeen states (Arizona, California, Colorado, Florida, Idaho, Indiana, Iowa, Kansas, Kentucky, Missouri, Montana, Oregon, South Dakota, Virginia, Washington, West Virginia, and Wisconsin) plus the District of Columbia started lotteries. Six more states (Georgia, Louisiana, Minnesota, Nebraska, New Mexico, and Texas) started lotteries during the 1990s. They were joined after 2000 by Arkansas, North Carolina, North Dakota, Oklahoma, South Carolina, and Tennessee. In 2013 Wyoming became the 44th state to institute a lottery.

LOTTERY GAMES

Early lottery games were simple raffles in which a person purchased a ticket that was preprinted with a number. The player typically had to wait for weeks for a drawing to determine if the ticket was a winner. These types of games, called passive drawing games, were the dominant lottery games in 1973. By 1997 they had ceased to exist, as consumers demanded more exciting games that provided quicker payoffs and more betting options.

Nearly all states that operate lotteries offer cash lotto and scratch-off instant games. Players of lotto games select a group of numbers from a large set and are then awarded prizes based on how many of their numbers are picked in a random drawing. Most lotto tickets sell for $1, and drawings are held once or twice per week to determine the winning numbers. Scratch-off instant games are paper tickets on which certain spaces have been coated with a scratch-off substance that when removed reveals numbers or text underneath that must match posted sequences to win.

Most states offer other numbers games, such as three- and four-digit games. Pull tabs, spiel, keno, and video lottery games are much less common. Pull tabs are two-ply paper tickets that must be separated to reveal symbols or numbers underneath that must match posted sequences to win. Spiel is an add-on feature to a lotto game that provides an extra set of numbers for a fee that must be matched to numbers selected in the random drawing to win. Keno is a lotto game in which a set of numbers is selected from a large field of numbers; players select a smaller set of numbers and are awarded prizes based on how many of their numbers match those in the drawn set. Video lottery terminals are electronic games of chance played on video screens that simulate popular casino games such as poker and blackjack. Keno and video lottery games are considered by many to be casino-type games, especially because they can be played every few minutes (in the case of fast keno) or at will (in the case of video lottery terminals), which makes them more controversial and generally less acceptable than more traditional lottery games.

As of 2017, many lottery games were conducted using computer networks. Retail outlets have computer terminals that are linked by phone lines to a central computer at the lottery commission, which records wagers as they are made. The computer network is a private, dedicated network that can be accessed only by lottery officials and retailers. Players can either choose their numbers themselves or allow the computer to select numbers randomly, an option known as Quick Pick. The computer link allows retailers to validate winning tickets.

Most lotto drawings are televised live. Some states also air lottery game shows in which contestants compete for money and prizes. For example, *The Big Spin*, the California State Lottery's 30-minute game show, was broadcast from 1985 to 2009. Michigan Lottery aired the quarterly game show *Make Me Rich!* from 2009 to 2012. Contestants mailed in instant lotto tickets to win a chance to compete in the show. Several games give contestants the chance to win a new car or up to $2 million.

Lottery winners generally have six months to one year to collect their prizes, depending on state rules. If the top prize, typically called the jackpot, is not won, the amount of the jackpot usually rolls over to the next drawing, increasing the jackpot. Lotteries are often most popular when the jackpot has rolled over several times and grown to an unusually large amount. Most states allow players to choose in advance how a jackpot will be paid to them—either all at once (the cash lump-sum prize) or in installments (an annuity, usually paid out over 20 or 25 years). Either way, in most states taxes are subtracted from the prize.

Vanchai Ariyabuddhiphongs of Bangkok University observes in "Lottery Gambling: A Review" (*Journal of Gambling Studies*, vol. 27, no. 1, March 2011) that lotteries are distinguished from other forms of gambling by two essential features. First, the probability of winning is extremely low, typically one in several million for modest prizes and one in several hundred million for big jackpots. The second distinguishing feature of lotteries is that they pay out a much lower ratio of the money collected than do casinos and other gambling operators. Ariyabuddhiphongs explains that "the payout ratio for lottery is typically 50%, compared to 74% in bingo, 81% in horseracing, 89% in slot machine, and 98% in blackjack played according to the basic rules."

Scratch Games

In 1974 Massachusetts became the first state to offer an instant lottery game using scratch-off tickets. By 2017 games involving scratch-off tickets (or "scratchers," as they are called in some states) were extremely popular. Lottery organizations offer many different scratch games with various themes.

Scratch games run for a specified period, usually for several months to a year. Many scratch-off tickets allow a player to win multiple times on each ticket. In 2017 most states offered "high-profit point tickets"—scratch-off tickets priced as high as $50, which are often part of a holiday or themed promotion. (Traditional scratch-off tickets sell for $1 to $5.) The higher-priced tickets appeal to many scratch players primarily because they offer higher payouts, in some cases exceeding $1 million.

Some scratch games offer prizes besides money, including merchandise, trips, vehicles, and tickets to sporting events or concerts. For example, in 2010 the Illinois Lottery unveiled a scratch game in which the top-five prizes were full tuition and fees for four years at the college or university of the winner's choice. Most states also offer "second-chance drawings," in which losing tickets in the main scratch-off game can be entered into a second contest. For example, the Texas Lottery offered a Dallas Cowboys–themed scratch-off in 2014 that allowed holders of losing tickets to enter periodic second-chance drawings whose top prizes included a package valued at $66,665, which granted the winner and 19 guests free airfare, hotel accommodations, a luxury stadium suite, and other insider perks for a Cowboys home football game. The total winnings for nonmonetary prizes often include payment by the lottery commission of federal and state income taxes on the value of the prizes.

Most lotteries operate toll-free numbers and websites that provide information on scratch-game prizes. Patrons can find out which prizes have been awarded and which remain to be claimed.

Powerball and Mega Millions

During the 1980s lottery officials realized that multistate lotteries could offer higher payoffs than single-state lotteries because the costs of running one game could be shared. The Multi-State Lottery Association (MUSL) was formed in 1987 as a nonprofit association of states offering lotteries. It administers a variety of games, the best known of which is Powerball. In this lotto game each ticket has six numbers: five numbers are selected out of 69 numbers, and then a separate number, the Powerball, is selected out of 26 numbers. A rival to Powerball, called the Big Game, was unveiled in 1996 in six states and expanded in the following years. It was renamed Mega Millions in 2002. In Mega Millions, players choose six numbers from two separate number pools: five numbers from 1 to 75, and one number from 1 to 15. All six numbers must be chosen in the drawing to win the jackpot. In 2010 an agreement between MUSL and Mega Millions allowed all state lotteries to begin selling both products, and in 2012 Mega Millions changed the structure of its game to allow for bigger jackpots. Powerball and Mega Millions jackpots routinely run to several hundreds of millions of dollars, and smaller prizes in each drawing range from a single dollar (the same as the cost of a ticket) to $5 million. The odds of winning a Powerball jackpot are approximately 1 in 292.2 million, and the odds of winning a Mega Millions jackpot are approximately 1 in 258.9 million.

As of August 2017, both Powerball and Mega Millions were offered in 44 states, the District of Columbia, and the U.S. Virgin Islands. The largest-ever Powerball jackpot, drawn on January 13, 2016, was valued at $1.6 billion, and the largest-ever Mega Millions jackpot, drawn on March 30, 2012, was valued at $656 million. These values refer to the jackpot when distributed in the form of an annuity (which gains value over time as a result of interest); when winners choose to receive their prizes in cash, the jackpot amounts are considerably lower. For example, the cash value of the $656 million Mega Millions jackpot was $474 million.

Other multistate lottery games are also available. Players of Hot Lotto compete for a $1 million jackpot by picking five white numbers from 1 to 47 and one orange number from 1 to 19. Numbers are drawn twice per week; if no one wins, the jackpot continues to grow. The game is available in 16 locations: Delaware, Idaho, Iowa, Kansas, Maine, Minnesota, Montana, New Hampshire, New Mexico, North Dakota, Oklahoma, South Dakota, Tennessee, Vermont, West Virginia, and the District of Columbia. Until 2016, Wild Card was offered in four states: Idaho, Montana, North Dakota, and South Dakota. Players picked five numbers between 1 and 31 and one of 16 different "wild cards." Drawings were held twice per week. 2by2 is played daily in Kansas, Nebraska, and North Dakota; players pick two red numbers and two white numbers between 1 and 26, giving them eight ways to win the $22,000 jackpot. MegaHits, a video lottery game, is available at video lottery terminals in Delaware, Maryland, Ohio, Rhode Island, and West Virginia.

HOW LOTTERIES OPERATE

In 2017 most lotteries were administered directly by state lottery boards or commissions. States are legally barred from fully privatizing their lotteries, but quasi-governmental or semi-privatized lottery corporations operate the lotteries in some states, including Connecticut, Georgia, Kentucky, Louisiana, and Tennessee. In 2011 Illinois became the first state to contract with an outside management firm to oversee its lottery's day-to-day operations, and since then Indiana, New Jersey, and Pennsylvania have followed suit. By outsourcing management to private businesses, these states hope to increase the profitability levels of their lotteries, although the funds are still earmarked for government programs. In most states enforcement authority regarding fraud and abuse rests with the attorney general's office, the state police, or the lottery commission. The amount of oversight and control that each legislature has over its lottery agency differs from state to state.

Although lotteries are a multimillion-dollar business, lottery commissions employ only a few thousand people nationwide. Lottery commissions set up, monitor, and run the games that are offered in their state, but the vast majority of lottery sales are made by retailers who contract to sell the games.

Retailers

According to the North American Association of State and Provincial Lotteries, in "Frequently Asked Questions" (2017, http://www.naspl.org/faq), lottery tickets were sold at more than 210,000 locations throughout North America in 2017. Most lottery ticket retailers were located in convenience stores, supermarkets, and gas stations.

Retailers get commissions on lottery sales and bonuses when they sell winning tickets. They also get increased store traffic and media attention, especially if they become known as "lucky" places to purchase lottery tickets. Some state lottery websites list the stores where winners purchased their tickets. For example, in "Historical Lucky Retailers" (2017, http://www.calottery.com/lucky-retailers/lucky-retailers-list), the California State Lottery notes that one retailer in Port Hueneme, California, had sold six winning million-dollar-plus tickets as of 2013; another retailer had sold four winning million-dollar-plus tickets, while six additional retailers in the state had each sold three.

Lottery tickets are often impulse purchases, so retailers sell them near the checkout. This also allows store operators to keep an eye on ticket vending machines to prevent play by underage customers. Because convenience stores increasingly offer pay-at-the-pump gasoline sales—transactions that are likely to decrease in-store traffic—some state lotteries have begun experimenting with selling lottery tickets at gas pumps and on the Internet.

In 2010 the Minnesota Lottery began allowing online purchases of Powerball and Mega Millions tickets, and in 2012 it became the first state to offer lottery sales at gas pumps. However, during the winter of 2014, when the lottery's executive director approved the sale of scratch-offs (which had the potential to increase off-site sales greatly), both houses of the state legislature passed legislation that would ban lottery sales online and at gas pumps. Amy Forliti reports in "Dayton Vetoes Bill Dumping Instant-Play Online Lottery Games, Bar Ticket Sales at Gas Pumps" (Associated Press, May 31, 2014) that Mark Dayton (1947–), the governor of Minnesota, vetoed the bill, arguing that the lottery's director was acting well within his authority by deciding to allow sales in these additional channels. According to Forliti, the state's tribal casino operators are staunchly opposed to allowing online and gas-pump lottery sales. Others opposed to the expansion are gas station owners who are worried that, with lottery tickets and gas available outside the store, even fewer customers will come inside, depressing sales of snacks, drinks, and other goods.

In 2011 Illinois also began offering online lottery sales, and as of 2017 it allowed Illinois residents over the age of 18 years to set up an online account to play Powerball, Mega Millions, and Lotto. However, scratch-off tickets still had to be purchased from brick-and-mortar retailers. The Missouri Lottery introduced ticket sales at a limited number of gas pumps and automatic teller machines in late 2013 and early 2014, and the California State Lottery followed suit in August 2014.

LOTTERY PLAYERS

Playing the lottery is by far the most popular form of gambling in the United States. In *2013 State of the States: The AGA Survey of Casino Entertainment* (2013, https://www.americangaming.org/sites/default/files/research_files/aga_sos2013_rev042014.pdf), the American Gaming Association indicates that 53% of its survey respondents purchased lottery tickets in 2012. By comparison, 32% gambled in a casino, 26% gambled casually with friends, 12% played poker, 6% placed bets on a race, and 3% gambled on the Internet.

Demographics

Ariyabuddhiphongs notes that in the United States, men play the lottery slightly more often than women, but that this is not true for all countries. Lottery participation does not vary markedly by age in the United States for those aged 60 years and younger. Those aged 61 years and older are less likely to play the lottery than younger people, but they spend more money on the lottery than other age groups.

Whereas the prevalence of gambling in general increases with education, studies consistently show that the prevalence of lottery play decreases with increasing levels of education, according to Ariyabuddhiphongs. Less-educated people tend to play the lottery more frequently than more-educated people, and countries with higher levels of education have lower lottery sales, on average, than countries with lower levels of education. Socioeconomic status is also highly correlated with the prevalence of lottery play. Ariyabuddhiphongs cites a 2002 U.S. study that examined the likelihood of lottery gambling among five quintiles divided by socioeconomic status. The highest rates of participation were found in the second- and third-lowest quintiles, and the lowest participation rates were found in the lowest and two highest quintiles. A 2006 study in Thailand found that almost two-thirds of those who purchased lottery tickets earned 25% less than the country's average monthly wages.

The fact that the most likely people to play lotteries are less-educated people of low socioeconomic status has led to much criticism of state lotteries and their marketing. Many scholars, journalists, and advocates for the poor contend that lotteries represent a regressive tax (a progressive tax is one in which richer people pay a larger share of their income than poorer people; a regressive tax is one in which poorer people pay a larger share of their income than richer people) of approximately 50% (the payout rate that a lottery player can expect over time). Although lotteries fund schools and other worthy public endeavors in all U.S. states, they do so, according to their critics, by taking money from those who can least afford to lose it. According to Ariyabuddhiphongs, "Sales data of lotteries from 39 states over 10 years showed a strong and positive relationship between sales and poverty rates.... Lottery sales increased with jackpot size and the increase was larger in areas with more economically disadvantaged populations."

TEXAS. Many state lottery commissions conduct their own demographic studies of lottery players, largely because they want to better target them in marketing campaigns. These studies, such as those published annually by the Texas Lottery Commission, provide valuable insights into lottery players.

In *Demographic Survey of Texas Lottery Players 2016* (December 2016, http://www.uh.edu/class/hobby/cpp/white-paper-series/_docs/HSPA%20White%20Paper%20Series_No.%208.pdf), the Texas Lottery Commission uses a random sample of 1,685 adult Texans to derive demographic information about lottery players compared with nonplayers and to estimate overall participation and the frequency of lottery purchases. As Table 7.2 shows, 589 survey respondents (35% of the total sample) reported past-year lottery play in 2016. This percentage was notably higher than those recorded in 2014 and 2015, when 25% and 28.7%, respectively, of respondents reported having played the lottery during the previous year. This rate, however, was substantially lower than the mid-1990s, when 71% of respondents reported past-year lottery play. (See Figure 7.1.)

The demographic profile of players was, in most respects, similar to that of nonplayers. Players and nonplayers had statistically similar profiles in terms of income, home ownership, marital and parenting status, race, education, and occupation. Notable statistical differences between players and nonplayers concerned employment status (46.8% of lottery players were employed full time, compared with 33.1% of nonplayers), age (those aged 55 years and older were significantly overrepresented among players), and marital status (married respondents were significantly overrepresented compared with those who had never been married). (See Table 7.2.)

Demographic differences were correlated much more strongly with the median monthly spending (the middle value; half spent less money and half spent more money) of lottery players, in keeping with the sizable body of research suggesting that lotteries disproportionately take from the poor and disadvantaged. The median amount of money spent per month for all lottery players surveyed was $13.00 in 2016. (See Table 7.3.) For Asian American respondents, however, the median monthly spending was $86.50, or more than six times greater than the average, while Native American respondents spent $49.00 a month, or nearly four times the median. Respondents aged 25 to

TABLE 7.2

Demographic characteristics of Texas lottery players vs. nonplayers, 2014–16

	Number and percentage responding		
Demographic factors	**All (sample size = 1,685)**	**Past-year players (sample size = 589)**	**Non-players (sample size = 1,096)**
Year[a]			
2016	1,685 (100%)	589 (35.0%)	1,096 (65.0%)
2015	1,979 (100%)	568 (28.7%)	1,411 (71.3%)
2014	1,701 (100%)	425 (25.0%)	1,276 (75.0%)
Income	**sample size = 818 (100%)**	**sample size = 346 (100%)**	**sample size = 472 (100%)**
Less than $12,000	62 (7.6%)	21 (6.1%)	41 (8.7%)
Between $12,000 and $19,999	51 (6.2%)	22 (6.4%)	29 (6.1%)
Between $20,000 and $29,999	61 (7.5%)	27 (7.8%)	34 (7.2%)
Between $30,000 and $39,999	69 (8.4%)	27 (7.8%)	42 (8.9%)
Between $40,000 and $49,999	70 (8.6%)	29 (8.4%)	41 (8.7%)
Between $50,000 and $59,999	76 (9.3%)	30 (8.7%)	46 (9.8%)
Between $60,000 and $74,999	94 (11.5%)	44 (12.7%)	50 (10.6%)
Between $75,000 and $100,000	117 (14.3%)	53 (15.3%)	64 (13.6%)
More than $100,000	218 (26.7%)	93 (26.9%)	125 (26.5%)
Employment status[c]	**sample size = 1,652 (100%)**	**sample size = 579 (100%)**	**sample size = 1,073 (100%)**
Employed full-time	626 (37.9%)	271 (46.8%)	355 (33.1%)
Employed part-time	109 (6.6%)	34 (5.9%)	75 (7.0%)
Unemployed/looking for work	63 (3.8%)	11 (1.9%)	52 (4.9%)
Not in labor force	105 (6.4%)	36 (6.2%)	69 (6.4%)
Retired	749 (45.3%)	227 (39.2%)	522 (48.7%)
Own or rent home[c]	**sample size = 1,607 (100%)**	**sample size = 564 (100%)**	**sample size = 1,043 (100%)**
Own	1,252 (77.9%)	455 (80.7%)	797 (76.4%)
Rent	284 (17.7%)	96 (17.0%)	188 (18.0%)
Occupied without payment	71 (4.4%)	13 (2.3%)	58 (5.6%)
Age of respondent[c]	**sample size = 1,332 (100%)**	**sample size = 469 (100%)**	**sample size = 863 (100%)**
18 to 24	61 (4.6%)	13 (2.8%)	48 (5.6%)
25 to 34	107 (8.0%)	50 (10.7%)	57 (6.6%)
35 to 44	120 (9.0%)	52 (11.1%)	68 (7.9%)
45 to 54	148 (11.1%)	70 (14.9%)	78 (9.0%)
55 to 64	284 (21.3%)	112 (23.9%)	172 (19.9%)
65 and over	612 (46.0%)	172 (36.7%)	440 (51.0%)
Marital status	**sample size = 1,622 (100%)**	**sample size = 568 (100%)**	**sample size = 1,054 (100%)**
Married	971 (59.9%)	353 (62.2%)	618 (58.6%)
Widowed	208 (12.8%)	50 (8.8%)	158 (15.0%)
Divorced	162 (10.0%)	66 (11.6%)	96 (9.1%)
Separated	22 (1.4%)	9 (1.6%)	13 (1.2%)
Never married	259 (16.0%)	90 (15.9%)	169 (16.0%)
Children under 18 living in household	**sample size = 1,530 (100%)**	**sample size = 538 (100%)**	**sample size = 992 (100%)**
Yes	281 (18.4%)	119 (22.1%)	162 (16.3%)
No	1,249 (81.6%)	419 (77.9%)	830 (83.7%)
Number of children under 18 living in household[d]	**sample size = 281 (100%)**	**sample size = 119 (100%)**	**sample size = 162 (100%)**
1	121 (43.1%)	48 (40.3%)	73 (45.1%)
2	94 (33.5%)	37 (31.1%)	57 (35.2%)
3	40 (14.2%)	21 (17.7%)	19 (11.7%)
4 or more	26 (9.3%)	13 (10.9%)	13 (8.0%)
Gender of respondent[d]	**sample size = 1,670 (100%)**	**sample size = 582 (100%)**	**sample size = 1,088 (100%)**
Male	756 (45.3%)	293 (50.3%)	463 (42.6%)
Female	914 (54.7%)	289 (49.7%)	625 (57.4%)
Race	**sample size = 1,585 (100%)**	**sample size = 555 (100%)**	**sample size = 1,030 (100%)**
White	1,132 (71.4%)	374 (67.4%)	758 (73.6%)
Hispanic	201 (12.7%)	83 (15.0%)	118 (11.5%)
African American	158 (10.0%)	75 (13.5%)	83 (8.1%)
Asian	26 (1.6%)	6 (1.1%)	20 (1.9%)
Native American Indian	23 (1.5%)	6 (1.1%)	17 (1.7%)
Other	45 (2.8%)	11 (2.0%)	34 (3.3%)
Hispanic origin	**sample size = 1,603 (100%)**	**sample size = 554 (100%)**	**sample size = 1,049 (100%)**
Yes	228 (14.2%)	95 (17.2%)	133 (12.7%)
No	1,375 (85.8%)	459 (82.9%)	916 (87.3%)

34 years spent $24.50 a month on average playing the lottery, more than any other age group. The monthly median spending for men ($16.00) was one-third higher than that for women ($12.00).

TABLE 7.2

Demographic characteristics of Texas lottery players vs. nonplayers, 2014–16 [CONTINUED]

Demographic factors	Number and percentage responding		
	All (sample size = 1,979)	Past-year players (sample size = 568)	Non-players (sample size = 1,411)
Education[b]	**sample size = 108 (100%)**	**sample size = 31 (100%)**	**sample size = 77 (100%)**
Less than high school	5 (4.6%)	0 (0.0%)	5 (6.5%)
High school graduate/GED	38 (35.2%)	10 (32.2%)	28 (36.4%)
Some college, no degree	27 (25.0%)	10 (32.3%)	17 (22.1%)
College degree	30 (27.8%)	10 (32.3%)	20 (26.0%)
Graduate/professional degree	8 (7.4%)	1 (3.2%)	7 (9.1%)
Occupation	**sample size = 1,119 (100%)**	**sample size = 409 (100%)**	**sample size = 710 (100%)**
Professional specialty	380 (34.0%)	117 (28.6%)	263 (37.0%)
Executive, administrative, and managerial	187 (16.7%)	70 (17.1%)	117 (16.5%)
Sales	110 (9.8%)	41 (10.0%)	69 (9.7%)
Service	108 (9.7%)	31 (7.6%)	77 (10.9%)
Administrative support, clerical	104 (9.3%)	43 (10.5%)	61 (8.6%)
Technicians and related support	81 (7.2%)	38 (9.3%)	43 (6.1%)
Machine operators, assemblers, and inspectors	29 (2.6%)	15 (3.7%)	14 (2.0%)
Private household	25 (2.2%)	8 (2.0%)	17 (2.4%)
Transportation and material moving	22 (2.0%)	12 (2.9%)	10 (1.4%)
Equipment handlers, cleaners, helpers, and laborers	19 (1.7%)	7 (1.7%)	12 (1.7%)
Farming, forestry, fishing	17 (1.5%)	6 (1.5%)	11 (1.6%)
Protective service	16 (1.4%)	10 (2.4%)	6 (0.9%)
Armed forces	14 (1.3%)	8 (2.0%)	6 (0.9%)
Precision productions, craft, and repair	7 (0.6%)	3 (0.7%)	4 (0.6%)

[a]There was an increase in the proportion of respondents who reported that they participated in any of the Texas Lottery games during the past year in 2016 from those who reported that they participated in 2015. The difference was statistically significant.
[b]Only those respondents who indicated that they were enrolled in school as full-time or part-time student were asked the question on education in the 2016 survey. We reported the percentage played and median dollars spent by education in the table. However, readers are cautioned that the number of responses in some sub-categories was too small (five or fewer) to provide statistically meaningful information.
[c]$^{**}p < 0.01$, $^{***}p < 0.001$, two-tailed test. There were statistically significant differences between players and non-players regarding the distribution by employment status, own or rent home, and age of the respondents.
[d]$^{*}p < 0.05$, $^{**}p < 0.01$, two-tailed test. There were statistically significant differences between players and non-players regarding the distribution by number of children under 18 living in household, gender, and Hispanic origin of the respondents.

SOURCE: "Table 2. Demographics: Summary for Income, Employment, Home Ownership, and Age," in *Demographic Survey of Texas Lottery Players 2016*, University of Houston, Hobby School of Public Affairs, December 2016, http://www.uh.edu/class/hobby/cpp/white-paper-series/_docs/HSPA%20White%20Paper%20Series_No.%208.pdf (accessed June 29, 2017). Courtesy of the Texas Lottery Commission.

Group Play

Groups of people frequently pool their money to buy lottery tickets, particularly for large jackpots. Group wins are beneficial to the lotteries because they generate more media coverage than solo wins and expose a wider group of friends, relatives, and coworkers to the idea that lotteries are winnable. Nonetheless, pooling arrangements, even those between only two people, can lead to disagreements if a group actually wins a jackpot. Several such groups have ended up in court, but given the number of winners every year, such cases are relatively rare.

Some states have formalized group play. For example, the California State Lottery started the Jackpot Captain program in 2001 to help "group leaders" manage lotto pools. Lotto captains have access to a special website that gives them tips on organizing and running group play. They can download and print forms that help them track players, games, dates, and jackpots. Lotto captains can also participate in special drawings for cash and prizes.

Why Do People Play Lotteries?

Ariyabuddhiphongs observes that the extremely low odds and the low payout ratio of lotteries make them objectively a losing proposition for players. Given that lotteries pay out 50 cents on the dollar, playing the lottery can be compared with paying a 50% tax. Over time, typical lottery players win small amounts to recoup the costs of purchasing tickets, but the likelihood of ever recouping more than around 50% is remote. The irrationality of purchasing lottery tickets is clear, and yet people willingly participate in lotteries worldwide in huge numbers. Why people play the lottery has been a question that researchers have been attempting to answer for decades. Although there remains no definitive answer, there are many provisional ones.

Some of the reasons people play lotteries resemble the reasons they partake in other gambling activities, according to Ariyabuddhiphongs's survey of research into the subject. A number of studies indicate that even mathematically inclined people and those with significant

FIGURE 7.1

Percentage of people who played any form of Texas lottery game in the past year, 1993–2016

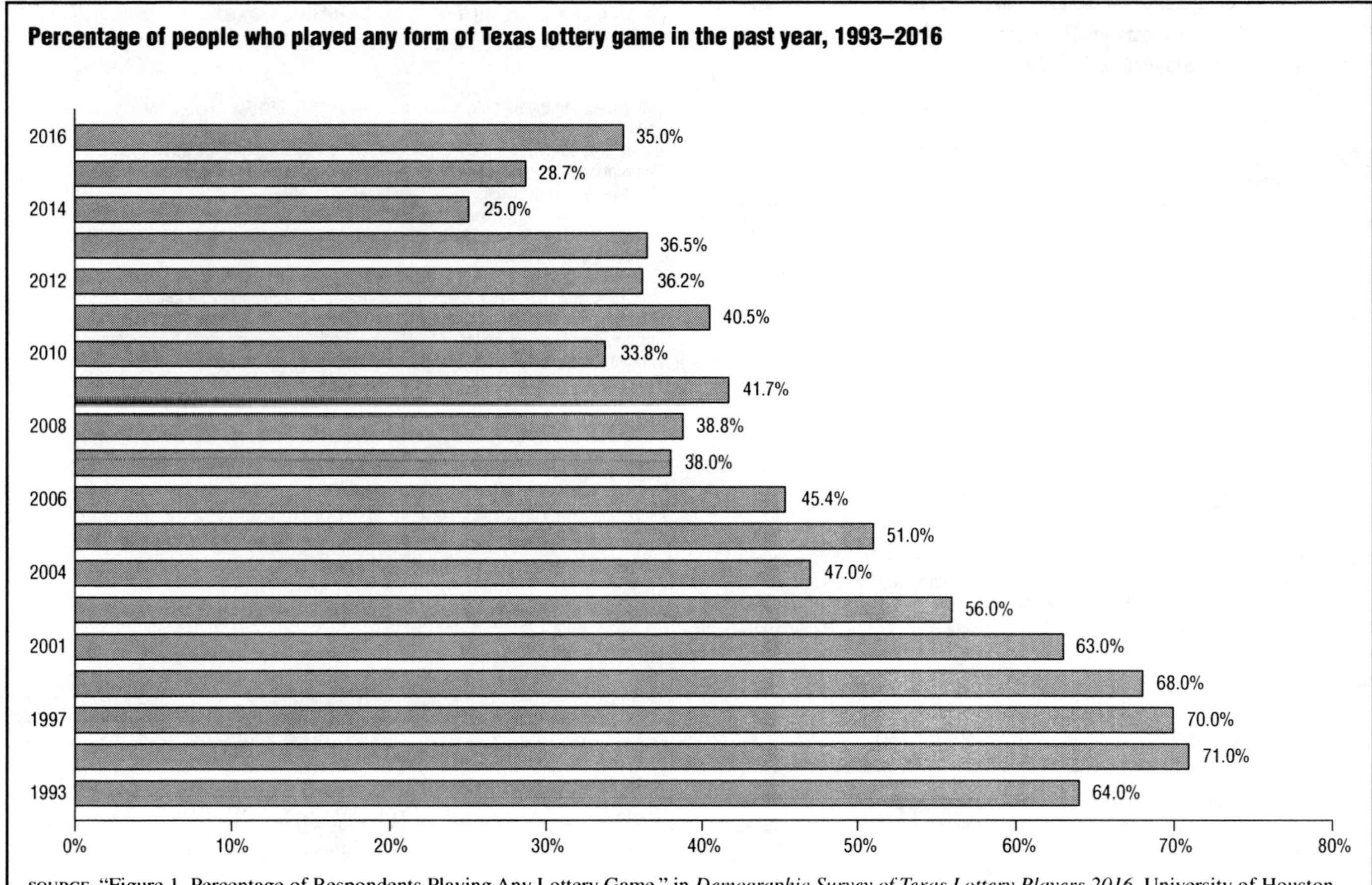

SOURCE: "Figure 1. Percentage of Respondents Playing Any Lottery Game," in *Demographic Survey of Texas Lottery Players 2016*, University of Houston, Hobby School of Public Affairs, December 2016, http://www.uh.edu/class/hobby/cpp/white-paper-series/_docs/HSPA%20White%20Paper%20Series_No.%208.pdf (accessed June 29, 2017). Courtesy of the Texas Lottery Commission.

knowledge and skill of gambling odds simply "switch off" their capacity for rational thought when they gamble. Accordingly, lottery players are subject to many of the irrational beliefs that are characteristic of casino gamblers, such as the gambler's fallacy and an unrealistic belief in one's own luck. Among the most commonly cited reasons for playing the lottery are the hope of winning a jackpot, which may be both cause and effect of an optimistic outlook on life; the need for money; and the sense of belonging that playing provides when friends engage in it together. Additionally, studies find that, in the United States as well as in other countries, lotteries are so commonplace that many people do not view them as a form of gambling. In any event, moral disapproval of gambling has waned with the rise of both lotteries and casinos in the United States, and many players consider gambling a positive influence on their lives and the lives of others.

THE ECONOMIC AND SOCIAL EFFECTS OF LOTTERIES

Lottery proponents usually use economic arguments to justify their position. They point out that lotteries provide state governments with a relatively easy way to increase their revenues without imposing more taxes. The games are financially beneficial to the many small businesses that sell lottery tickets and to the larger companies that participate in merchandising campaigns or that provide advertising or computer services. In addition, lottery advocates suggest that the games provide cheap entertainment to people who want to play, while raising money for the betterment of all.

Lottery opponents also have economic arguments. They contend that lotteries contribute only a small percentage of total state revenues and, therefore, have a limited effect on state programs. Lotteries cost money to operate, and they use false hopes to lure people into parting with their money. In addition, as mentioned earlier, opponents contend that the marketing and business strategies of lotteries unfairly target those with little education and financial resources and that the net effect is similar to a regressive tax aimed at the poor.

The Division of Lottery Money

In statistics relating to lottery proceeds, the amount of money categorized as sales represents the total amount taken in by the lottery. This sales amount is then split between prizes, administrative costs, retailer commissions,

TABLE 7.3

Median spending on Texas state lottery tickets, by demographic characteristics of players, 2014–16

Year	Percentage played	Median dollars spent
2016[a] (overall population = 1,685)	35.0 (sample = 589)	$13.00
2015 (overall population = 1,979)	28.7 (sample = 568)	$10.00
2014 (overall population = 1,701)	25.0 (sample = 425)	12.00
Demographic factors 2016		
Education[b]		
Less than high school diploma	—[c]	—
High school diploma (sample = 38)	26.3 (sample = 10)	15.50
Some college (sample = 27)	37.0 (sample = 10)	18.00
College degree (sample = 30)	33.3 (sample = 10)	18.00
Graduate degree	—	—
Income		
Under $12,000 (sample = 62)	33.9 (sample = 21)	20.00
$12,000 to $19,999 (sample = 51)	43.1 (sample = 22)	14.50
$20,000 to $29,999 (sample = 61)	44.2 (sample = 27)	26.00
$30,000 to $39,999 (sample = 69)	39.1 (sample = 27)	24.00
$40,000 to $49,999 (sample = 70)	41.4 (sample = 29)	12.00
$50,000 to $59,999 (sample = 76)	39.5 (sample = 30)	15.00
$60,000 to $74,999 (sample = 94)	46.8 (sample = 44)	9.00
$75,000 to $100,000 (sample = 117)	45.3 (sample = 53)	20.00
More than $100,000 (sample = 218)	42.7 (sample = 93)	10.00
Race		
White (sample = 1,132)	33.0 (sample = 374)	10.00
African American (sample = 158)	47.5 (sample =75)	17.00
Hispanic (sample = 201)	41.3 (sample = 83)	22.00
Asian (sample = 26)	23.1 (sample = 6)	86.50
Native American Indian (sample = 23)	26.1 (sample = 6)	49.00
Other (sample = 45)	24.4 (sample = 11)	13.00
Hispanic origin		
Yes (sample = 228)	41.7 (sample = 95)	24.00
No (sample = 1,375)	33.4 (sample = 459)	10.00
Gender		
Female (sample = 914)	31.6 (sample = 289)	12.00
Male (sample = 756)	38.8 (sample = 293)	16.00
Age		
18 to 24 (sample = 61)	21.3 (sample = 13)	3.00
25 to 34 (sample = 107)	46.7 (sample = 50)	24.50
35 to 44 (sample = 120)	43.3 (sample = 52)	17.50
45 to 54 (sample = 148)	47.3 (sample = 70)	13.50
55 to 64 (sample = 284)	39.4 (sample = 112)	11.50
65 or older (sample = 612)	28.1 (sample = 172)	13.00
Employment status		
Employed full/part time (sample = 735)	41.5 (sample = 305)	16.00
Unemployed (sample = 63)	17.5 (sample = 11)	—
Retired (sample = 749)	30.3 (sample = 227)	11.00

TABLE 7.3

Median spending on Texas state lottery tickets, by demographic characteristics of players, 2014–16 [CONTINUED]

[a]The increase in the participation rates from 2015 to 2016 was statistically significant.
[b]Only those respondents who indicated that they were enrolled in school as full-time or part-time student were asked the question on education in the 2016 survey. We reported the percentage played and median dollar spent by education in the table but no statistical tests were conducted on them.
[c]There were only five or fewer respondents in this sub-category and therefore it was not reported. The reporting rule was used for all subsequent tables in the report unless otherwise stated.
Note: Hispanic origin $p < 0.05$, gender and age < 0.01, employment status < 0.001. The significance notations refer only to the "percentage played" column. In some categories, the number of respondents contributing to cell percentages is small. This small size has the effect of making generalizations from these figures more tenuous. Due to greater uncertainty, small sample size also requires larger discrepancies among categories to attain acceptable levels of statistical significance. We note in the discussion of individual lottery games those instances where sub-samples are especially small.

SOURCE: "Table 3. Any Game: Past-Year Lottery Play and Median Dollars Spent per Month by Demographics," in *Demographic Survey of Texas Lottery Players 2016*, University of Houston, Hobby School of Public Affairs, December 2016, http://www.uh.edu/class/hobby/cpp/white-paper-series/_docs/HSPA%20White%20Paper%20Series_No.%208.pdf (accessed June 29, 2017). Courtesy of the Texas Lottery Commission.

and state profits. In general, 50% to 60% of U.S. lottery sales are paid out as prizes to winners. Administrative costs for advertising, employee salaries, and other operating expenses usually account for 1% to 10% of sales. On average, retailers collect 5% to 8% of sales in the form of commissions and approximately 2% as bonuses for selling winning tickets. The remaining 30% to 40% is turned over to the state.

As Table 7.1 shows, U.S. lotteries generated $66.8 billion in sales in 2015, and $42.3 billion was distributed to prize winners. Another $3.2 billion was spent on administrative costs, which includes payments to retailers. The states collectively took in $21.4 billion in profits, to be used on various government programs.

Table 7.4 shows how the states spent their fiscal year (FY) 2016 lottery proceeds. (The dollar amounts in Table 7.4, which come from the North American Association of State and Provincial Lotteries, differ slightly from those in Table 7.1, which come from the Census Bureau.) The most common use of state lottery proceeds is the funding of public education. Many of the top-earning lotteries directed their profits exclusively to education, including those in California, Florida, Georgia, and New York. Other lotteries dedicated exclusively or mainly to funding education included Arkansas, Idaho, Kentucky, Michigan, Missouri, New Hampshire, New Jersey, New Mexico, North Carolina, Oklahoma, South Carolina, Tennessee, Texas, Vermont, Virginia, and West Virginia. Other lotteries earmarked most or all of their proceeds to the state's general fund, the pool of money that typically accounts for most state spending and that primarily funds education, public assistance, Medicaid, and corrections. Lotteries directing some portion of proceeds to the general fund included those in Arizona, Connecticut, Delaware, the District of Columbia, Iowa, Maine, Maryland, Massachusetts, Michigan, Minnesota, Montana, North Dakota, Rhode Island, South Dakota, and Washington. Colorado directed portions of its lottery proceeds to the Division of Parks and Wildlife, a Conservation Trust Fund, and the Great Outdoors Colorado Trust Fund; Maine, Minnesota, Nebraska, and Oregon directed portions of their proceeds to similar programs. Meanwhile, Wisconsin earmarked its lottery proceeds to property tax relief.

Unclaimed Lottery Winnings

Unclaimed lottery winnings add up to hundreds of millions of dollars each year, and each state handles them

TABLE 7.4

Lottery contributions to beneficiaries, by state, fiscal year 2016

State (with beneficiaries)	Total contributions (in millions)
Arizona	**$205.69**
Heritage Fund	
Commerce Authority Arizona Competes Fund	
Mass transit	
Healthy Arizona	
General Fund (by category)	
Court-Appointed Special Advocate Fund (unclaimed prizes)	
Homeless shelters	
Department of Gaming	
University Bond Fund	
Internet Crimes Against Children Victims	
Tribal College Dual Enrollment Program	
Arkansas	**$85.27**
Educational Trust Fund	
California	**$1,590.00**
Education	
Colorado	**$143.57**
Division of Parks and Wildlife	
Conservation Trust Fund	
Great Outdoors Colorado Trust Fund	
School Fund	
Connecticut	**$339.80**
General Fund (to benefit education, roads, health and hospitals, public safety, etc.)	
Delaware	**$213.06**
General Fund	
Health and Social Services-Problem Gamblers Programs	
District of Columbia	**$53.17**
General Fund	
Florida	**$1,700.00**
Education Enhancement Trust Fund	
Georgia	**$1,100.00**
Lottery for Education Account	
Idaho	**$49.50**
Public Schools (K-12)	
Public Buildings	
Illinois	**$694.39**
Illinois Common School Fund	
Illinois Veterans Assistance Fund	
Ticket for the Cure Fund	
Quality of Life Endowment Fund	
Multiple Sclerosis Research Fund	
Special Olympics Fund	
Indiana	**$281.47**
Build Indiana Fund	
Teachers Pension Fund	
Police & Firefighters Pension Fund	
Iowa	**$88.02**
General Fund	
Veterans Trust Fund	
Kansas	**$78.21**
Transfers to the state	
Kentucky	**$253.04**
Post-Secondary & College Scholarships	
Literacy Programs & Early Childhood Reading	
Louisiana	**$177.93**
Transfers to state	
Problem Gambling	

TABLE 7.4

Lottery contributions to beneficiaries, by state, fiscal year 2016 [CONTINUED]

State (with beneficiaries)	Total contributions (in millions)
Maine	**$57.58**
General Fund	
Maine Outdoor Heritage Fund	
Maryland	**$1,030.00**
Maryland General Fund—Lottery Profit	
Baltimore City Schools—Lottery Profit	
Maryland Stadium Authority—Lottery Profit	
Education Trust Fund—VLT and Table Game Proceeds	
Local Impact Grants—VLT Proceeds	
Horse Racing Industry—VLT Proceeds	
Small, Minority, and Women-Owned Businesses—VLT Proceeds	
Responsible Gaming—VLT and Table Game Assessments	
Massachusetts	**$989.41**
Cities and towns	
Arts Council	
General Fund	
Compulsive Gamblers	
Michigan	**$889.90**
Education (K-12)	
Health and Human Services	
General Fund	
Minnesota	**$144.68**
General Fund	
Environmental and Natural Resources Trust Fund	
Game & Fish Fund	
Natural Resources Fund	
Compulsive Gambling	
Missouri	**$302.58**
Public Education in Missouri	
Montana	**$12.89**
State of Montana General Fund	
Nebraska	**$42.78**
Compulsive Gamblers Assistance Fund	
Education Innovation Fund	
Environmental Trust Fund	
State Fair Support & Improvement Fund	
Nebraska Opportunity Grant Fund	
New Hampshire	**$79.18**
Education	
New Jersey	**$987.00**
Education and Institutions	
New Mexico	**$46.30**
Lottery Tuition Fund	
New York	**$3,300.00**
Education	
North Carolina	**$635.27**
Education	
Alcohol Law Enforcement	
NC Problem Gambling	
North Dakota	**$10.32**
Compulsive Gambling Fund	
State General Fund	
Drug Task Force	
Ohio	**$1,116.00**
Education	
Oklahoma	**$67.1**
Education	
Mental Health	

differently. Some states, such as New York, require that unclaimed winnings be returned to the prize pool. Other states allocate such funds to lottery administrative costs or to specific state programs. For example, in Texas

TABLE 7.4

Lottery contributions to beneficiaries, by state, fiscal year 2016 [CONTINUED]

State (with beneficiaries)	Total contributions (in millions)
Oregon	**$572.93**
Economic Development Public Education Parks and Natural Resource Programs Gambling Addiction Prevention & Treatment Programs	
Pennsylvania	**$1,121.00**
Older Pennsylvanians	
Rhode Island	**$369.76**
General Fund	
South Carolina	**$398.91**
Education Lottery Fund	
South Dakota	**$117.65**
General Fund Capital Construction Fund Grant to Human Services	
Tennessee	**$394.05**
Lottery for Education Account After School Program	
Texas	**$1,390.00**
Foundation School Fund Multicategorical Teaching Hospital Texas Veterans Commission (Veterans Assistance Fund)	
Vermont	**$26.41**
Education Fund	
Virginia	**$588.19**
Direct Aid to Public Education K-12	
Washington	**$175.52**
Washington Opportunity Pathways Account King County Stadium and Exhibition Center (Qwest Field) Economic Development Strategic Reserve Problem Gambling General Fund	
West Virginia	**$496.67**
Education Senior citizens Tourism Other	
Wisconsin	**$158.45**
Public Benefit—Total available for property tax relief	
Wyoming	**$2.04**
Wyoming Cities, Town, and Counties	

SOURCE: Created by Stephen Meyer for Gale, © 2017. Data from the North American Association of State and Provincial Lotteries.

unclaimed prizes go to funds that benefit hospital research and payment of indigent health care. The California State Lottery reports in "Unclaimed Prizes" (2017, http://www.calottery.com/sitecore/content/ARCHIVE/media/fact-sheets/unclaimed-prizes) that as of FY 2013–14 almost $793.5 million in unclaimed money had been allocated to school funding.

Taxes and Other Withholding from Lottery Winnings

Lottery winnings are usually taxable as personal income. All prizes greater than $600 are reported by the lotteries to the Internal Revenue Service. In general, the lottery agencies subtract taxes before awarding large prizes. Besides federal tax withholding, lottery winners must pay all applicable state and local taxes, as with any form of income. Thus, winners who play and win their home state's lottery effectively contribute twice to their state government. Winners of multistate jackpots such as Powerball and Mega Millions pay income taxes on their winnings in the state where they reside.

Effects of Lottery Funding on Education

Lottery proponents often justify lotteries by noting their substantial ability to generate funding for education. Indeed, most states donate some lottery proceeds to education, and many donate all proceeds to education. For example, the California State Lottery, which exclusively funds public education, contributed nearly $1.6 billion to education in FY 2016. (See Table 7.4.)

Nevertheless, $1.6 billion is not a large amount in the context of California's overall budget. As Table 7.5 shows, California had total 2015 revenues (including taxes, payments from the federal government, and other forms of revenue) of $332.5 billion, and the state's total spending on education was $90.5 billion. Proceeds from the California State Lottery thus represented approximately 0.5% of total state revenues and 1.7% of total education spending.

Opponents further argue that, in many cases, lottery funds do not represent additional dollars for education but simply replace general fund dollars that would have been spent on education anyway. In "Mega Millions: Do Lotteries Really Benefit Public Schools?" (WashingtonPost.com, March 30, 2012), Valerie Strauss notes that many states typically respond to pressures to decrease expenditures by cutting education budgets and then simply using lottery proceeds to maintain the funding they have cut. Similarly, Pat Garofalo reports in "Mega Failure: Why Lotteries Are a Bad Bet for State Budgets" (ThinkProgress.org, March 30, 2012) that although the funds that lotteries generate sometimes meet a state's short-term budget needs, lottery sales tend to reach a ceiling due to market saturation and competition from other states, whereas state spending necessarily climbs due to population growth, inflation, and other factors.

In "Texas Lottery: A Different Game Than State Was Sold Two Decades Ago" (Statesman.com, September 7, 2010), Eric Dexheimer wonders whether the Texas Lottery is worth the cost it imposes on the state's people, in the form of extracting money from those who can least afford it. Dexheimer notes that a smaller percentage of Texans played the lottery in 2010 (40%) than in 1994 (70%). Thus, lottery marketers had to convince those who continued to play the lottery to spend more money just to keep pace with budgetary expectations. Meanwhile, although most of the state's lottery proceeds go

TABLE 7.5

California government finances, 2015

[Amounts in thousands of dollars]

Item	California
Total revenue	**332,456,557**
General revenue	265,498,842
Intergovernmental revenue	82,907,118
Taxes	151,234,165
General sales	38,464,704
Selective sales	13,947,655
License taxes	9,420,863
Individual income tax	77,929,551
Corporate income tax	9,007,182
Other taxes	2,464,210
Current charges	21,373,082
Miscellaneous general revenue	9,984,477
Utility revenue	1,002,780
Liquor stores revenue	—
Insurance trust revenue	65,954,935
Total expenditure	**330,502,626**
Intergovernmental expenditure	97,968,655
Direct expenditure	232,533,971
Current operation	152,273,413
Capital outlay	10,183,475
Insurance benefits and repayments	57,750,839
Assistance and subsidies	4,930,538
Interest on debt	7,395,706
Exhibit: Salaries and wages	32,021,699
Total expenditure	**330,502,626**
General expenditure	270,064,716
Intergovernmental expenditure	97,968,655
Direct expenditure	172,096,061
General expenditure, by function:	
Education	90,546,647
Public welfare	109,031,702
Hospitals	10,633,226
Health	8,648,877
Highways	10,529,860
Police protection	1,790,672
Correction	9,667,904
Natural resources	3,814,057
Parks and recreation	529,286
Governmental administration	7,485,654
Interest on general debt	7,168,706
Other and unallocable	10,218,125
Utility expenditure	2,687,071
Liquor stores expenditure	—
Insurance trust expenditure	57,750,839
Debt at end of fiscal year	151,715,007
Cash and security holdings	845,232,818

SOURCE: Adapted from "Summary Table," in *2015 Annual Survey of State Government Finances*, U.S. Census Bureau, 2017, https://www2.census.gov/govs/state/2015_summary_table.xls (accessed June 29, 2017)

to education, lottery funds as a percentage of education spending have been steadily shrinking. "In 1996," Dexheimer explains, "lottery proceeds paid for about two weeks of schooling for Texas students. [In 2010] the money raised by the lottery will barely cover three days."

Such arguments are consistent with the earlier findings of Donald E. Miller of Saint Mary's College, who argues in "Schools Lose Out in Lotteries" (USAToday.com, April 15, 2004) that educational spending per student gradually decreases once a state starts a lottery. He examines data for 12 states that had enacted lotteries for education between 1965 and 1990 and finds that, before lotteries were established, the average education spending in those states increased each year by approximately $12 per student. In the first year following the initiation of the lotteries, the states increased their education spending on average by nearly $50 per student. Nevertheless, the increase fell sharply in subsequent years and eventually lagged behind states without lottery-generated education funds.

Compulsive Gambling and Cognitive Distortion

The vast majority of states operate lotteries, and as a result lotteries are easily accessible to large numbers of people. Surveys show that lottery play is the most popular and widely practiced form of gambling in the United States. However, because most states with lotteries also offer commercial or tribal casino gambling, it is difficult to assess lotteries' role in contributing to gambling disorders. Does the combination of easy and widespread access and general public acceptance mean that lottery players are more likely to develop serious gambling problems?

Dean Gerstein et al. conclude in the landmark study *Gambling Impact and Behavior Study: Report to the National Gambling Impact Study Commission* (April 1, 1999, http://govinfo.library.unt.edu/ngisc/reports/gibstdy.pdf) that there is a significant association between lottery availability and the prevalence of at-risk gambling within a state. At-risk gamblers are defined as those who gamble regularly and may be prone to a gambling problem. However, the researchers find that multivisit lottery patrons had the lowest prevalence of pathological and problem gambling among the gambling types examined.

Ariyabuddhiphongs's 2011 review of the scholarly literature on lotteries suggests that the introduction of lotteries results in a consequent increase in the number of people who fit the criteria for pathological gambling in the fourth edition of the American Psychiatric Association's *Diagnostic and Statistical Manual of Mental Disorders* (*DSM-IV*; these criteria were revised and reclassified under the term *gambling disorder* for the 2013 edition of *DSM-5*). Nevertheless, the extreme popularity of lottery play relative to other forms of gambling and the ready availability of lottery tickets in most states have not been accompanied by a proportional elevation in the percentage of the population estimated to be pathological or disordered gamblers. As is discussed in Chapter 6, the overall prevalence of disordered gambling is generally regarded as being 0.2% and 2.1%.

In "The States' Role in Gambling Addiction" (NYTimes.com, November 4, 2010), Dr. Timothy Fong of the University of California, Los Angeles, answers *New York Times* readers' questions about problem gambling on the newspaper's *Consults* medical blog. He explains, "State lotteries come in a variety of different formats and games. There is no scientific data that I have seen that shows that lottery play can lead directly to gambling addiction. Playing the lottery is still gambling, though, so lottery games that offer a high reward with high frequency—for example, 'scratch' games or video lottery terminals—probably carry a higher chance of harm."

THE FUTURE OF U.S. LOTTERIES

New State Lotteries?

As of August 2017, only six states did not have lotteries: Alabama, Alaska, Hawaii, Mississippi, Nevada, and Utah. Hawaii and Utah do not permit any types of gambling and seem unlikely to amend their constitutions. A lottery in Nevada is very unlikely because of the tremendous growth of casino gambling there. Alaskan politicians have shown minimal interest in a lottery. Although many state lottery bills have been introduced in the Alabama and Mississippi legislatures, most of them died in committee, and the rest were soundly defeated on the floor.

As of 2017, the most recent state to have instituted a lottery was Wyoming. A bill to establish a state lottery was passed in 2013, and the Wyoming Lottery began operating in mid-2014. Prior to Wyoming, the most recently established state lotteries were those of Arkansas, which began operating in 2009, and North Dakota, which began operating in 2004.

Jackpot Fatigue

A major problem facing the lottery industry is called "jackpot fatigue." Lottery consumers demand higher and higher jackpots so they can stay excited about lotto games. However, individual states cannot increase jackpot sizes without either greatly increasing sales or decreasing the portion of lottery revenue going to public funds. The first option is difficult to achieve and the second is politically dangerous. Jackpot fatigue has driven increasing membership in multistate lotteries, such as Powerball and Mega Millions.

Pressure for Increased Revenue

Besides coping with jackpot fatigue, many lotteries also face pressure to increase the amount of profit going to government programs. Several states are considering decreasing their lottery payout to raise much needed funds. Opponents argue that cutting prize payouts will reduce sales, thereby making it nearly impossible to increase state revenues.

Growth in lottery revenues stagnated after double-digit growth in the last two decades of the 20th century. The public perception is that lotteries significantly contribute to state budgets and shrink residents' taxes. However, as noted earlier in the discussion of California's state budget, the reality is that lotteries account for no more than a small percentage of state budgets. Possibilities for future growth in lottery revenues include pay-at-the-pump and online lottery sales. States have also been attempting to introduce higher-priced scratch-off lottery tickets in an attempt to boost sluggish revenues. Karen Farkas observes in "Ohio Lottery Considering Offering a $50 Instant Ticket" (Cleveland.com, April 25, 2014) that Ohio debuted a $30 scratch-off lottery ticket in January 2014, and between January and April the new high-priced ticket accounted for $71.9 million in sales, or 17% of total scratch-off sales. Because of the $30 ticket's success, the state's lottery officials subsequently introduced a $50 ticket, the "Hit $50!" By 2017 a number of other state lotteries had begun offering scratch tickets at that price point.

CHAPTER 8
SPORTS GAMBLING

Wagering on sporting events is one of the oldest and most popular forms of gambling in the world. The ancient Romans gambled on chariot races, animal fights, and gladiator contests. The Romans brought sports and gambling to Britain during the first century AD, where they have flourished for hundreds of years. Cockfighting, bear- and bullbaiting, wrestling, and footraces were popular sporting events for gambling throughout Europe during the 16th and 17th centuries. Horse races and boxing matches became popular spectator and betting sports during the 18th century. During the 19th and 20th centuries sporting events became more team-oriented and organized as rugby, soccer, and cricket grew in popularity.

Many early colonists who traveled to North America brought their love of sports and gambling with them. Horse racing, in particular, became a part of American culture. However, the morals of the late 18th and early 19th centuries constrained popular support for legalized sports gambling. By 1910 almost all forms of gambling were illegal in the United States. This did not stop people from gambling on sports, however. The practice continued to flourish, and horse racing, in particular, managed to maintain some legal respectability as a betting sport.

Nevada legalized gambling again in 1931, in no small part as a response to the Great Depression (1929–1939). Other states legalized horse racing during the 1930s, similarly motivated by the need to generate economic growth at a time of mass unemployment. Meanwhile, for two decades Nevada permitted wagering on college and professional sports beyond horse racing. Point-shaving scandals in college basketball and the exposure of the gambling industry's connections to organized crime by a federal investigation led to a crackdown during the 1950s. Legal sports gambling did not return to Nevada until 1975, when it was tightly licensed and regulated. In 1992, at the urging of Paul Tagliabue (1940–), the commissioner of the National Football League (NFL), Congress passed the Professional and Amateur Sports Protection Act, which limited legal sports betting to only four states: Delaware, Montana, Nevada, and Oregon.

Between the 1930s and the 1970s horse racing flourished as a spectator sport and as a forum for gamblers. Outside of Nevada, racetracks had a monopoly on the gambling business until the late 1960s, when the first states began introducing lotteries. The rise of lotteries and, later, of commercial and tribal casinos marginalized racetrack gambling, which by the 1990s represented only a fraction of the overall U.S. gaming industry.

In the 21st century sports gambling in the United States can be broken down into three primary categories: pari-mutuel betting (in which those who have bet on the top competitors split the pool of winnings) on events such as horse and greyhound races and on the ball game jai alai (a court game played only in Florida, in which players bounce a ball against a wall and catch it using a long curved basket strapped to the wrist); legal betting using a bookmaker in Nevada; and illegal betting using a bookmaker or the Internet.

THE PREVALENCE OF SPORTS GAMBLING

Betting on sports through legal channels such as racetracks and Nevada bookmakers is one of the least popular forms of gambling in the United States. The Harris Poll indicates in "Pro Football Is Still America's Favorite Sport" (January 26, 2016, http://www.theharrispoll.com/sports/Americas_Fav_Sport_2016.html) that only 1% of Americans surveyed regarded horse racing to be their favorite sport in 2015. This figure was roughly comparable to the popularity of track and field (2%) and behind boxing (3%), men's golf (3%), and men's soccer (4%). Similarly, betting on sporting events through a bookmaker in Nevada is a niche offering. As Table 8.1 shows, sports betting generated $219.2 million for the Nevada gaming industry

TABLE 8.1

Nevada sports betting financial data, 1984–2016

[Amounts in thousands of dollars]

	Units	Win amount	% change	Win%	Drop	Total gaming win	% total	Win/Unit
1984	51	20,899	—	2.34%	894,564	3,092,980	0.68%	409.78
1985	56	21,485	2.80%	2.52%	851,786	3,309,372	0.65%	383.66
1986	59	34,921	62.54%	1.00%	1,081,161	3,492,474	1.00%	591.88
1987	67	29,054	−16.80%	0.74%	1,875,400	3,949,427	0.74%	433.64
1988	66	42,023	44.64%	3.19%	1,318,052	4,429,023	0.95%	636.71
1989	71	45,776	8.93%	3.35%	1,366,448	4,750,716	0.96%	644.73
1990	90	48,839	6.69%	0.89%	1,671,479	5,480,664	0.89%	542.66
1991	104	52,300	7.09%	2.80%	1,867,857	5,579,128	0.94%	502.88
1992	114	50,602	−3.25%	2.81%	1,800,783	5,864,228	0.86%	443.88
1993	112	75,035	48.28%	3.74%	2,006,283	6,247,508	1.20%	669.96
1994	112	122,450	63.19%	5.73%	2,136,998	7,007,586	1.75%	1,093.30
1995	119	79,415	−35.14%	3.27%	2,428,693	7,368,580	1.08%	667.35
1996	133	76,486	−3.69%	3.08%	2,483,312	7,426,192	1.03%	575.08
1997	136	89,718	17.30%	3.69%	2,431,382	7,801,920	1.15%	659.69
1998	146	77,375	−13.76%	3.41%	2,269,062	8,064,970	0.96%	529.97
1999	156	109,192	41.12%	4.42%	2,470,407	9,021,570	1.21%	699.95
2000	165	123,836	13.41%	5.33%	2,323,377	9,602,586	1.29%	750.52
2001	162	118,077	−4.65%	5.78%	2,042,855	9,468,559	1.25%	728.87
2002	149	110,383	−6.52%	5.70%	1,936,544	9,447,660	1.17%	740.83
2003	150	122,630	11.10%	6.58%	1,863,678	9,625,304	1.27%	817.53
2004	159	112,504	−8.26%	5.39%	2,087,273	10,562,247	1.07%	707.57
2005	164	126,176	12.15%	5.59%	2,257,174	11,649,040	1.08%	769.37
2006	168	191,538	51.80%	7.89%	2,427,605	12,622,044	1.52%	1,140.11
2007	169	168,363	−12.10%	6.49%	2,594,191	12,849,137	1.31%	996.23
2008	177	136,441	−18.96%	5.29%	2,579,225	11,599,124	1.18%	770.85
2009	185	136,380	−0.04%	5.31%	2,568,362	10,392,675	1.31%	737.19
2010	183	151,096	10.79%	5.47%	2,762,267	10,404,731	1.45%	825.66
2011	183	140,731	−6.86%	4.89%	2,877,935	10,700,994	1.32%	769.02
2012	182	170,062	20.84%	4.93%	3,449,533	10,860,715	1.57%	934.41
2013	186	202,838	19.27%	5.60%	3,622,107	11,142,915	1.82%	1,090.53
2014	190	227,045	11.93%	5.82%	3,901,117	11,018,688	2.06%	1,194.97
2015	196	231,787	2.09%	5.47%	4,237,422	11,114,081	2.09%	1,182.59
2016	192	219,174	−5.44%	4.86%	4,509,753	11,257,147	1.95%	1,141.53

SOURCE: David G. Schwartz, "Sports Betting Win, 1984–2016," in *Nevada Sports Betting Totals: 198–2016*, University of Nevada, Las Vegas, Center for Gaming Research, 2017, http://gaming.unlv.edu/reports/NV_sportsbetting.pdf (accessed June 30, 2017)

in 2016; this figure represented only 2% of the total gaming win of $11.3 billion reported by the state that year. (See Figure 4.1 in Chapter 4.)

In "Sports Gambling in U.S.: Too Prevalent to Remain Illegal?" (WashingtonPost.com, February 27, 2015), Will Hobson reports that the amount spent on legal sports wagering in the United States is dwarfed by the money devoted to illegal sports gambling. Although precise figures are difficult to determine, Hobson estimates that gamblers spend between $80 billion and $380 billion on illegal sports betting each year. The predominance of illegal sports betting ultimately prompted Adam Silver (1962–), the commissioner of the National Basketball Association, to publish an influential op-ed piece in the *New York Times*. In "Legalize and Regulate Sports Betting" (NYTimes.com, November 13, 2014), he calls for a nationwide expansion of legalized sports gambling. David Purdum writes in "GAME Act Proposing Repeal of Federal Prohibition on Sports Betting Revealed" (ESPN.com, May 25, 2017) that the U.S. House of Representatives, Energy and Commerce Committee issued in May 2017 the Gaming Accountability and Modernization Enhancement (GAME) Act, a proposed law that would repeal the Professional and Amateur Sports Protection Act, thereby opening the door for legalized sports betting in individual states. As of August 2017, the GAME Act had not yet come before Congress for a vote.

PARI-MUTUEL GAMBLING

Pari-mutuel is a French term meaning "mutual stake." In pari-mutuel betting all wagers on a particular event or race are combined into a pool that is split between the winning bettors, minus a percentage for the management; consequently, the larger the pool, the bigger the payoff. In pari-mutuel gambling patrons bet against each other, not against the house. The principles of the pari-mutuel system were developed in France by Pierre Oller during the late 19th century.

The pari-mutuel system has been used for horse races in the United States since about 1875, but it did not really catch on until the 1920s and 1930s, when the totalizator, an automatic odds calculator, came into use. The totalizator took money, printed betting tickets, and continuously calculated odds based on betting volume.

Previously, horse betting had been conducted mostly by bookmakers who were notoriously corrupt. In 1933 California, Michigan, New Hampshire, and Ohio legalized pari-mutuel gambling on horse racing as a means of regulating the industry and gaining some revenue. Dozens of states followed suit over the next decade. Pari-mutuel gambling was also adopted for greyhound racing and jai alai matches. As of August 2017, 43 U.S. states allowed pari-mutuel gambling. A handful of states permit pari-mutuel gambling by law but do not have facilities or systems in place to conduct it.

In pari-mutuel gambling the entire amount wagered is called the betting pool, the gross wager, or the handle. The system ensures that event managers receive a share of the betting pool, regardless of who wins a particular race or match. The management's share is called the takeout. The takeout percentage is set by state law and is usually about 20%.

Because betting on horses is subject to a substantial degree of skill, this 20% takeout does not represent a strict house advantage. Unskilled pari-mutuel gamblers, however, should not expect more than an 80% return on their wagers over time. By comparison, games requiring no skill whatsoever, such as slot machines, offer gamblers roughly a 90% return on their money. This is one of the reasons that racetracks are less popular than casinos among the general population of gamblers.

Breakage refers to the odd cents that are not paid out to winning bettors because payoffs are rounded. For example, the payout on a $2 bet is typically rounded down in $0.10 or $0.20 increments, and the cents left over are the breakage. Although breakage amounts to only pennies per bet, it adds up quickly with high betting volume. The California Horse Racing Board notes in *46th Annual Report of the California Horse Racing Board* (2017, http://www.chrb.ca.gov/annual_reports/2016_annual_report.pdf) that California horse racetracks accumulated approximately $7.6 million in breakage during fiscal year (FY) 2015–16. Each state has its own rules about breakage, but usually the funds are split between the state, the track operators, and the winning horse owners. Breakage is subtracted from the betting pool before payouts are made.

Pari-mutuel wagering can be performed in person at the event or at offtrack betting (OTB) facilities. Some states also allow betting by telephone or by Internet when an account is set up before bet placement. Many races are broadcast as they occur by televised transmission to in-state and out-of-state locations (including OTB sites). This process, known as simulcasting, allows intertrack wagering to take place. In other words, bettors at one racetrack can place bets on races taking place at another racetrack.

A race book is an establishment (usually a room at a casino or racetrack) in which intertrack wagering takes place on pari-mutuel events such as horse and greyhound races. A race book typically features many television monitors that show races as they occur. Race books are included in many Nevada and Atlantic City, New Jersey, casinos as well as in some tribal casinos.

HORSE RACING

Horse racing has been a popular sport since the time of the ancient Greeks and Romans. It was popularized in western Europe during the Middle Ages, when knights returned from the Crusades with fast Arabian stallions. These horses were bred with sturdy English mares to produce a new line of horses now known as Thoroughbreds. Thoroughbreds are tall, lean horses with long, slender legs. They are renowned for their speed and grace while running.

Thoroughbred racing became popular among the British royalty and aristocrats, earning it the nickname the "sport of kings." The sport was transplanted to North America during colonial times. In *Thoroughly Thoroughbred* (2006, http://www.jockeyclub.com/pdfs/thoroughly_thoroughbred.pdf), the Jockey Club, the governing body of Thoroughbred horse racing in North America, indicates that races were run on Long Island, New York, as early as 1665. However, the advent of organized Thoroughbred racing in the United States is attributed to Governor Samuel Ogle (1702–1752) of Maryland, who staged a race between pedigreed horses in "the English style" in Annapolis, Maryland, in 1745. The Annapolis Jockey Club, which sponsored the race, later became the Maryland Jockey Club. Among its members were George Washington (1732–1799) and Thomas Jefferson (1743–1826).

Thoroughbred breeding was prominent in Maryland and Virginia until the Civil War (1861–1865), when many operations were moved to Kentucky. Thoroughbred racing had already grown popular throughout the agricultural South. In 1863 the Saratoga racecourse opened in northern New York. It is considered the oldest Thoroughbred flat track in the country. (A flat track is one with no hurdles or other obstacles for a racehorse to jump over.) The Jockey Club, which maintains the official breed registry for Thoroughbred horses in North America, was established in 1894 in New York City.

Horse racing remained popular in the United States until World War II (1939–1945), when it was severely curtailed. Although the popularity of horse racing declined further between the 1950s and 1970s, the pari-mutuel industry remained comparatively vibrant due to the absence of other gambling options in most U.S. states. Since the 1990s the underlying economics of horse racing have been eroded, chiefly by competition from lotteries and casinos but also, in the view of many analysts, by the proliferation of other forms of entertainment and leisure activities.

In response, racetrack operators have lobbied for the right to place slot machines and other gaming devices at tracks, with the goal of boosting their bottom lines and bolstering the prize purses available to the owners of winning horses. Such facilities are typically called "racinos." As of August 2017, racinos operated in 11 states: Delaware, Florida, Iowa, Louisiana, Maine, New Mexico, New York, Oklahoma, Pennsylvania, Rhode Island, and West Virginia.

Thoroughbred Racetracks and Races

The Daily Racing Form states in "Racing Links: Race Tracks" (2017, http://www.drf.com/racing_links/links_tracks.html) that 93 Thoroughbred racetracks of varying sizes operated throughout the United States in 2017. Some are open seasonally, whereas those in warm climates are open year-round. Some are owned by the government, and some are owned by public or private companies. Thoroughbred horse racing in the United States is controlled by a relatively small group of participants. Companies such as Churchill Downs Incorporated and the Stronach Group Inc. own multiple prominent racetracks, as does the nonprofit New York Racing Association. Analysts predict that the industry will continue to undergo consolidation, with corporations taking over most of the business.

The three most prestigious Thoroughbred races in the United States are the Kentucky Derby at the Churchill Downs racetrack in Kentucky, the Preakness Stakes at the Pimlico racetrack in Maryland, and the Belmont Stakes at the Belmont Park racetrack in New York. The races are held during a five-week period between May and June of each year. A horse that wins all three races in one year is said to have won the Triple Crown. As of August 2017, only 12 horses had captured the Triple Crown—the most recent was by a horse named American Pharoah in 2015.

According to the Jockey Club, in "Number of Races" (2017, http://www.jockeyclub.com/default.asp?section=FB&area=6), there were 37,614 Thoroughbred horse races in 2016. The number of Thoroughbred races held each year has declined steadily since 1989.

Non-Thoroughbred Horse Racing

Other types of horse racing also attract pari-mutuel wagering. In harness racing, specially trained horses trot or pace rather than gallop. Usually, the horse pulls a sulky (two-wheeled cart), which carries a jockey who controls the reins. Sometimes the jockey is seated on the horse rather than in the sulky. Harness racing is performed by standardbred horses, which are shorter, more muscled, and longer in body than Thoroughbreds. The National Association of Trotting Horse Breeders in America established the official registry for standardbred horses in 1879. At the time, Thoroughbred horses were the favorite of high society, and standardbred horses were popular among the common people. In "Track Information" (2017, http://www.ustrotting.com/trackside/trackfacts/trackfacts.cfm), the U.S. Trotting Association indicates that in 2017 there were 35 licensed harness racetracks that offered pari-mutuel betting.

A third type of horse known for racing is the quarter horse, so named because of its high speed over distances of less than a quarter of a mile. North American colonists originally bred quarter horses to be both hardworking and athletic. According to the American Quarter Horse Association, in "Race and Track Information" (2017, https://www.aqha.com/racing/pages/horseman-info/track-and-race-information/), there were 71 quarter-horse racing tracks in 2017, some of which were also used for Thoroughbred races and some of which were state fair sites.

Arabian horses are considered to be the only purebred horses in the race circuit. The Arabian Jockey Club indicates in "Race Meet Calendar 2017" (January 13, 2017, https://34z5as1d9gu01m1geu13gzkc-wpengine.netdna-ssl.com/wp-content/uploads/2017racemeet.pdf) that 16 Arabian races were scheduled for 2017.

Betting on Horse Races

The betting pool for a particular horse race depends on how much is wagered by bettors. Each wager affects the odds. The more money bet on a horse, the lower that horse's odds and the potential payoff become. The payout for winning tickets is determined by the amount of money bet on the winner in relation to the amount bet on all the other horses in that particular race.

First, the takeout is subtracted from the betting pool. This money goes toward track expenses, taxes, and the purse. Most states also require that a portion of the takeout be put into a breeder incentive fund to encourage horse breeding and health in the state. After the takeout and the breakage are subtracted from the betting pool, the remaining money is divided by the number of bettors to determine the payoff, or return, on each wager.

The odds on a particular horse winning first, second, or third place are estimated on the morning of a race and then are constantly recalculated by computer during the betting period before the race. The odds are posted on a tote board and on television screens throughout the betting area. The tote board also tallies the total amount paid into each pool. Bettors can wager that a particular horse will win (come in first), place (come in first or second), or show (come in first, second, or third). The payoff for a win is higher than payoffs for place or show because the latter two pools have to be split more ways. For example, the show pool must be split between all bettors who selected win, place, or show.

Betting on horse races is considered to be more a game of skill than a game of chance. Professional racing bettors spend many hours observing individual horses and consider previous race experience when they make their picks. This process, called "handicapping," gives them some advantage over bettors who pick a horse based on whim—for example, because they like its name. Although bettors do not play directly against each other, an individual bettor's skill does affect other bettors because the odds are based on the bets of all gamblers.

The Economic Effects of Horse Racing

HANDLE. Although live attendance at racetracks and the total number of races have long been in decline, the amount of money gambled on horse races continued to increase through 2003. Since then, the pari-mutuel handle for Thoroughbred races has fallen significantly. According to the Jockey Club, in "Pari-Mutuel Handle" (2017, http://www.jockeyclub.com/default.asp?section=FB&area=8), the pari-mutuel handle from Thoroughbred horse racing was approximately $10.7 billion in 2016, down 29% from the high of $15.2 billion in 2003. (See Table 8.2.) Approximately $9.7 billion (90%) of the total amount gambled on Thoroughbred races in 2016 was wagered at OTB facilities.

As noted earlier, racetrack operators typically enjoy a takeout of 20% of the total handle. Given consistent, industry-wide declines in the handle (which are even more severe than the previously noted figures suggest, once inflation is taken into account), traditional racetrack operators often struggle to break even in the 21st century. The decline in the fortunes of racetrack operators, which has been ongoing for decades, is one of the primary factors behind the push to add slot machines and other gaming options at U.S. racetracks. Besides generating steady profits for track owners, a substantial portion of racino gaming machine revenue typically goes to the racing purse. Thus, tracks with racinos have a significant competitive advantage over those that do not, in terms of both profitability and the ability to attract the owners of top Thoroughbreds to their races (because a primary factor that influences top owners' choice of races is the size of the purse).

Some of the most storied racetracks in the United States are located in states that do not allow them to operate as racinos, including those in Kentucky and California. Conor Dougherty reports in "Horse Racing's Slide Spurs New Bet on Track Land" (WSJ.com, May 2, 2014) that a track such as Churchill Downs, which hosts the world-famous Kentucky Derby, can cope with racing's declining profitability in the absence of racinos

TABLE 8.2

Total wager amounts, North American Thoroughbred racing, 1990–2016

[Dollars in millions]

					Canada*			Puerto Rico*			
Units	On track	Off track	Total	% change	On track	Off track	Total	On track	Off track	Total	Total*
1990	—	—	9,385	1.1	—	—	823	—	—	—	10,208
1991	—	—	9,393	0.1	—	—	804	—	—	—	10,198
1992	—	—	9,639	2.6	—	—	770	—	—	—	10,409
1993	—	—	9,600	−0.4	—	—	731	—	—	—	10,331
1994	—	—	9,897	3.1	—	—	681	—	—	—	10,578
1995	—	—	10,429	5.4	—	—	795	—	—	—	11,224
1996	2,944	8,683	11,627	11.5	259	383	642	10	257	267	12,536
1997	2,703	9,839	12,542	7.9	217	310	527	9	249	258	13,327
1998	2,498	10,617	13,115	4.6	188	310	498	7	185	192	13,805
1999	2,359	11,365	13,724	4.6	161	278	439	7	238	245	14,408
2000	2,270	12,051	14,321	4.4	150	325	475	10	236	246	15,042
2001	2,112	12,487	14,599	1.9	153	387	540	8	208	216	15,355
2002	2,029	13,033	15,062	3.2	153	414	567	8	221	229	15,858
2003	1,902	13,278	15,180	0.8	139	397	536	8	218	226	15,942
2004	1,860	13,239	15,099	−0.5	137	364	501	8	228	236	15,836
2005	1,741	12,820	14,561	−3.6	144	424	568	8	239	247	15,376
2006	1,688	13,097	14,785	1.5	109	419	528	8	229	237	15,550
2007	1,670	13,055	14,725	−0.4	132	375	507	8	206	214	15,446
2008	1,489	12,173	13,662	−7.2	96	393	489	7	188	195	14,346
2009	1,325	10,990	12,315	−9.9	87	403	490	6	162	168	12,973
2010	1,199	10,220	11,419	−7.3	78	425	503	5	149	154	12,076
2011	1,229	9,541	10,770	−5.7	73	449	522	5	140	145	11,437
2012	1,239	9,643	10,882	1.0	70	486	556	6	127	133	11,571
2013	1,185	9,692	10,877	−0.0	58	431	489	5	113	118	11,484
2014	1,175	9,377	10,552	−3.0	56	414	470	4	94	98	11,120
2015	1,122	9,553	10,675	1.2	47	470	517	4	94	98	11,290
2016	1,058	9,686	10,744	0.6	55	508	563	5	86	91	11,398

SOURCE: "Pari-Mutuel Handle," in *2017 Fact Book*, The Jockey Club, 2017, http://www.jockeyclub.com/default.asp?section=FB&area=8 (accessed June 30, 2017)

because its connection to the sport's marquee event substantially enhances its ability to generate revenue. In contrast, many other fabled tracks struggle to remain open. In 2014 the sites of two California tracks located on prime suburban real estate—Bay Meadows in the San Francisco Bay Area and Hollywood Park in the Los Angeles area (both of which had been in existence since the 1930s)—were being redeveloped as office parks and residential areas. Dougherty notes that prior to Bay Meadows's final racing season, in 2008, profits had fallen "from eight-figure levels in 1997 to barely above break-even."

PURSE. The prize money that the owners of horses receive for a successful race is called the purse. Historically, only the top four finishers in a Thoroughbred race received a portion of the purse, but since the 1970s different ways of dividing the purse have emerged, some of which earmark money for horses that finish fifth or lower, in the interest of encouraging participation. Different tracks and states utilize different formulas, but in most cases the owner of the first-place horse receives around 60% of the purse. The owner is responsible for paying the horse's trainer and jockey. The owners of the horses finishing second and third typically receive about 20% and 12%, respectively, of a race purse.

In "Gross Purses" (2017, http://www.jockeyclub.com/default.asp?section=FB&area=7), the Jockey Club notes that in 2016 the gross purse (the total amount of money awarded to owners) for U.S. Thoroughbred racing was $1.1 billion, roughly equal to the previous year's total. The annual gross purse in the United States remained relatively consistent between 2007 and 2016. (See Figure 8.1.)

BREEDING. The horse racing industry also supports a large business in horse breeding. In 1962 Maryland was the first state to establish a program to encourage breeders within the state through direct money payments. The practice spread quickly to other states involved in horse racing. According to the Jockey Club, in "Distribution of Registered US Foal Crop by State" (2017, http://www.jockeyclub.com/default.asp?section=FB&area=4), Kentucky was the leading Thoroughbred breeding state, responsible for 8,130 foals in 2015, or 38.7% of the total U.S. foal crop. Florida followed with 2,098 foals, or 10% of the total, while California (1,833 foals) was responsible for 8.7% of the national total. The total number of Thoroughbred foals born in the United States in 2015 was 21,025, down nearly 40% from the 2006 total of 34,905.

Thoroughbred horses are typically sold as yearlings. In "Average/Median Price Per Yearling" (2017, http://www.jockeyclub.com/default.asp?section=FB&area=14), the Jockey Club indicates that in 2016, 6,963 yearlings were sold in the United States, at an average price of $60,594 and a median price (the middle value; half of yearlings went for a lower price and half went for a higher price) of $18,000.

GREYHOUND RACING

Greyhounds are mentioned in many ancient documents. English noblemen used greyhounds to hunt rabbits, a sport known as coursing. Greyhound racing was given the nickname the "sport of queens" because Queen Elizabeth I (1533–1603) of England established the first formal rules for greyhound coursing during the 1500s. Greyhounds were brought to the United States during the

FIGURE 8.1

Total purse amounts, North American Thoroughbred racing, 2007–16

[Dollars in millions]

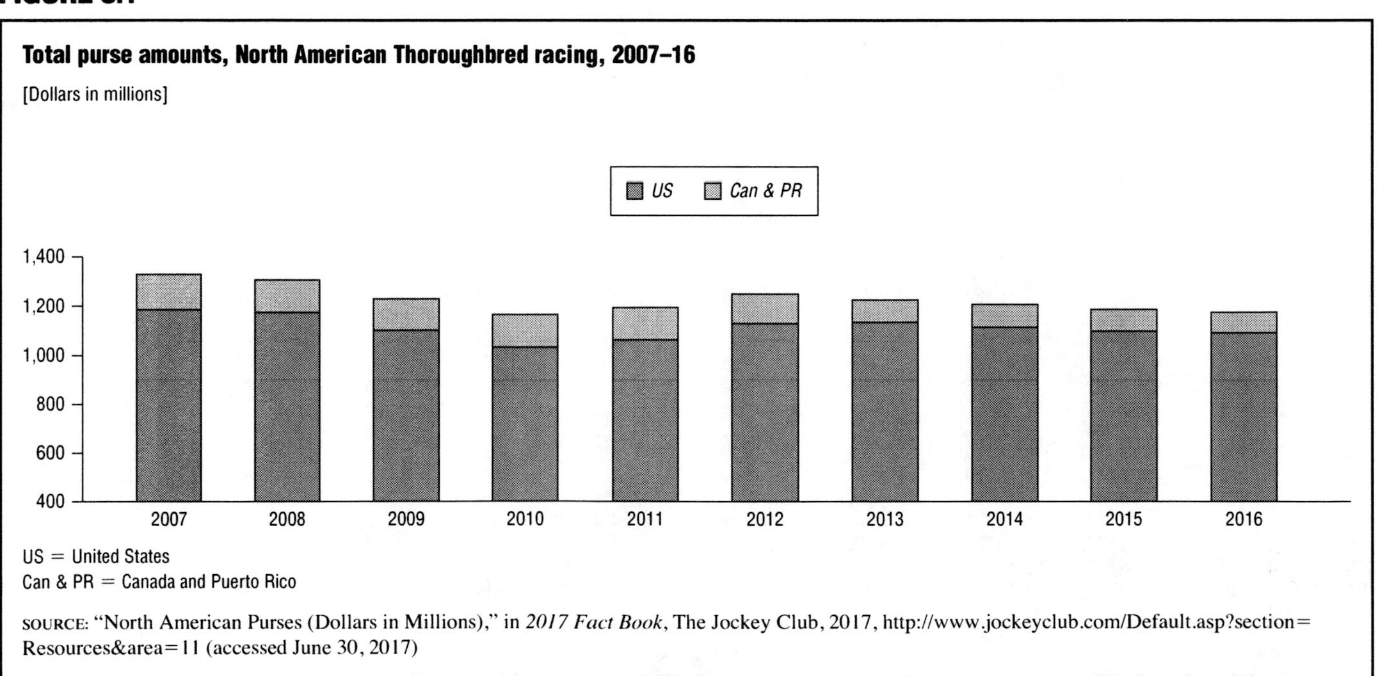

US = United States
Can & PR = Canada and Puerto Rico

SOURCE: "North American Purses (Dollars in Millions)," in *2017 Fact Book*, The Jockey Club, 2017, http://www.jockeyclub.com/Default.asp?section=Resources&area=11 (accessed June 30, 2017)

late 1800s to help control the jackrabbit population on farms in the Midwest. Eventually, farmers began holding local races, using live rabbits to lure the dogs to race. During the early 1900s Owen Patrick Smith (1867–1927) invented a mechanical lure for this purpose. The first circular greyhound track opened in Emeryville, California, in 1919.

Three major organizations manage greyhound racing in the United States: the National Greyhound Association (NGA), which represents greyhound owners and is the official registry for racing greyhounds; the American Greyhound Track Operators Association (AGTOA); and the American Greyhound Council, a joint effort of the NGA and the AGTOA, which manages the industry's animal welfare programs, including farm inspections and adoptions.

Wagering on greyhound races is similar to wagering on horse races. Greyhound racing, however, is not nearly as popular as horse racing, and its popularity has declined dramatically in the past few decades. Part of this decline in popularity is due to the same factors that have led to the decline in horse racing's popularity (i.e., competition from lotteries, casinos, and other forms of entertainment and leisure activities). Additionally, growing awareness of the costs imposed on the dogs themselves by the racing industry's standard operating procedures has led to widespread disapproval of dog racing. According to advocates for the dogs' welfare, hundreds of dogs at each racetrack are confined in small cages for most of each day, their well-being neglected as a matter of course. Serious injuries during races are common, the dogs are fed unsafe meat, and some trainers are suspected of injecting their dogs with performance-enhancing drugs that undermine the dogs' health. Additionally, when the dogs are no longer suitable for racing, which is usually around the age of four to six years, the standard practice is to kill them.

In response to this policy of killing retired racing dogs, movements promoting the adoption of former racing greyhounds have emerged to ameliorate the problem somewhat. Additionally, a number of advocacy organizations, including Grey2K USA Worldwide, People for the Ethical Treatment of Animals, and the American Society for the Prevention of Cruelty to Animals, have increasingly pushed for state legislation that outlaws greyhound racing. According to Grey2K USA Worldwide, in "State by State" (2017, http://www.grey2kusa.org/action/states.html), since the 1990s bans on dog racing have been passed by the state legislatures of Maine (1993), Virginia (1995), Vermont (1995), Idaho (1996), Washington (1996), Nevada (1997), North Carolina (1998), Pennsylvania (2004), Massachusetts (2010), New Hampshire (2010), Rhode Island (2010), Colorado (2014), and Arizona (2016). Since 2001, 31 tracks, or roughly half of the entire total number in operation, have either closed or ceased holding dog races. As of August 2017, pari-mutuel betting on greyhound racing was both legal and ongoing in only six states: Alabama, Arkansas, Florida, Iowa, Texas, and West Virginia.

Grey2K USA Worldwide reports in *Fact Sheet: Greyhound Racing in the United States* (August 21, 2017, https://www.grey2kusa.org/pdf/GREY2KUSANationalFactSheet.pdf) that the annual amount wagered on greyhound racing decreased 70% during the first decade and a half of the 21st century, falling from nearly $2 billion in 2001 to just over $500 million in 2014.

LEGAL SPORTS GAMBLING

Beyond pari-mutuel gambling, legalized sports gambling is extremely limited in the United States. As of 2017, only Nevada offered a full slate of options for gambling on sporting events. Delaware, Montana, and Oregon allowed limited forms of betting.

As discussed earlier in this chapter, in 1992 Congress passed the Professional and Amateur Sports Protection Act, which banned sports betting nationwide except in four states (Delaware, Montana, Nevada, and Oregon) that had offered it at some point between 1976 and 1990. Delaware was exempted from the law because it ran an NFL lottery in 1976 that allowed only parlay bets (bets in which the bettor had to pick the winners of at least three different NFL games in a single wager). In 2009 Delaware reintroduced gambling on NFL games and hoped to expand to other sports as a way to solve a state budget crisis that was caused by the Great Recession, which lasted from late 2007 to mid-2009. However, a federal appeals court ruled in August 2009 that Delaware's sports betting plan violated federal law and must be limited to betting that was allowed in 1976 (i.e., parlay bets on only the NFL). As a result of this ruling, Delaware's three slot parlors could take parlay bets only on NFL games.

In March 2011 a New Jersey ballot measure that would have legalized sports betting in the state passed by a wide margin, and in 2012 Governor Chris Christie (1962–) signed the resulting legislation that would have paved the way for his administration to begin issuing sports gambling licenses to casinos in Atlantic City as well as to the state's racetrack operators. In response, the nation's four major professional sports leagues—the NFL, Major League Baseball, the National Basketball Association, and the National Hockey League—joined the National Collegiate Athletic Association (NCAA) in a lawsuit filed against the state of New Jersey, asserting that the state's plan to legalize sports betting was in violation of federal law. In 2013 a federal court judge ruled against New Jersey, arguing that neither state legislation nor judicial decisions were the proper forum for overturning the Professional and Amateur Sports Protection Act. The only remedy for the people of New Jersey, the judge suggested, was for Congress to repeal or amend the law. Two subsequent federal court rulings sided with the leagues. John Brennan, however, reports in "U.S. Supreme Court to Review N.J. Sports Betting Case; Supporters React" (NorthJersey.com, June 27, 2017) that the U.S. Supreme Court agreed in June 2017 to hear the case, providing legalization advocates with another opportunity to present their arguments. The case was still pending as of August 2017.

Sports and Race Books in Nevada

Sports books are establishments that accept and pay off bets on sporting events. Bettors must be over the age of 21 years and be physically present in the state.

According to the Nevada Gaming Control Board, in *Gaming Revenue Report: December 31, 2016* (2017, http://gaming.nv.gov/modules/showdocument.aspx?documentid=11852), in 2016 there were 193 sports pools operating in the state, along with 137 race books. Most of these operations were based in Las Vegas casinos. The typical casino book is a large room with many television monitors showing races and games from around the world. Most casinos have combined race/sports books, although the betting formats are usually different. Race book betting is mostly of the pari-mutuel type, whereas sports book betting is by bookmaking.

Bookmaking

Bookmaking is the common term for the act of determining odds and receiving and paying off bets. The person performing the service is called the bookmaker or bookie. Bookmaking has its own lingo, which can be confusing to those who are not familiar with it. For example, a dollar bet is actually a $100 bet, a nickel bet is a $500 bet, and a dime bet is a $1,000 bet. To place a bet with a bookmaker, the bettor lays down (pays) a particular amount of money to win a particular payoff.

Bookmakers make money by charging a commission called "vigorish" (often shortened to just "vig"). Vigorish is an important yet misunderstood concept by most bettors. Most gambling literature describes vigorish as a 4.55% commission that a bookie earns from losers' bets. In "A Crash Course in Vigorish … and It's Not 4.55%" (2017, http://www.professionalgambler.com/vigorish.html), J. R. Martin, a sports handicapper (a person who analyzes betting odds and gives advice to bettors), provides a different interpretation of vigorish. Statistically, only bettors who win exactly half of their bets pay exactly 4.55% in vigorish. Other bettors pay different percentages. Martin explains that a bettor must win 53% of all equally sized bets to break even. This bettor, however, would wind up paying a vigorish of at least 4.82%.

Some sports bets are simple wagers based on yes or no logic. Examples include under and over bets, in which a bettor wagers that a particular game's final score will be under or over a specific number of points.

Most sports bets are based on a line that is set by the bookmaker. For example, the line for an NFL game between the Miami Dolphins and the Tennessee Titans might say that the Dolphins are picked by seven points. A bettor picking the Dolphins to win the game wins money only if the Dolphins win the game by more than seven points.

The line does not reflect a sports expert's assessment of the number of points by which a team will win. Rather, it is a concept designed to even up betting and to ensure that the bookmaker gets bets on both sides. This reduces the bookie's financial risk. Bookmakers will change lines if one side receives more betting action than the other. The skill of sports gambling comes in recognizing the accuracy of the line. Experienced bettors choose games in which they believe the posted lines do not accurately reflect the expected outcomes. This gives them an edge. The odds for most licensed sports books in Nevada are set by Las Vegas Sports Consultants, Inc.

Developments in Legal Sports Gambling in Nevada

Nevada legalized gambling during the Great Depression as a means of raising revenue. During that time, Charles K. McNeil (1903–1981), a Chicago securities analyst, developed the handicapping system, in which bookmakers establish the betting line. The new system provided an incentive for gamblers to bet on the underdog in a contest and made gambling more appealing. During the 1940s the Nevada legislature legalized off-track betting on horse racing, and sports and race books were popular in the state's casinos.

Then in 1950 and 1951 a series of U.S. Senate hearings led by Senator Estes Kefauver (1903–1963; D-TN) investigated the role of organized crime in the gambling industry. The televised hearings focused the nation's attention on gangsters, corrupt politicians, and legal and illegal gambling. One of the results was the passage of a 10% federal excise tax on "any wager with respect to a sports event or a contest." Because of the tax, the casino sports books, which were making only a small profit, were forced to shut down.

In 1974 the federal excise tax was reduced to 2%, and the sports books slowly made a comeback. Frank "Lefty" Rosenthal (1929–2008), a renowned handicapper, was credited with popularizing the sports book in Las Vegas during the 1970s. The 1980s were boom years for the sports and race books: in 1983 the federal excise tax was reduced to 0.3%. Jimmy Snyder (1919–1996) brought some legitimacy to sports gambling through his appearances on televised sports shows. The amount of money wagered in the Nevada sports books increased dramatically until the mid-1990s, when it began to level off.

Money and Games

Wagering in Nevada is allowed only on professional and college sports; it is not allowed on high school sporting events or Olympic events. Nevada law also restricts the sports books to wagering on events that are athletic contests—betting is not allowed on related events, such as who will win the most valuable player awards.

Table 8.3 provides a breakdown of the amounts wagered on sports betting in Nevada between 1992 and 2016. Nearly $1.7 billion was wagered on football games in 2016; this total represented more than one-third (37.5%) of the $4.5 billion bet on all sports that year. As Table 8.4 shows, betting on football also generates more revenues than any other sport. Nevertheless, football's share of overall sports betting revenues varies widely from year to year. In 1994 football wagers accounted for more than half (51.9%) of all Nevada sports betting revenues. (See Table 8.5.) By contrast, in 2005 revenues generated from football wagers represented just under 32% of the total sports betting win in Nevada; this figure was roughly comparable to that of basketball, which accounted for 30.4% of all sports betting revenues that year.

Low-Stakes Sports Gambling

MONTANA. Montana allows five types of sports gambling: sports pools, Calcutta pools, sports tab games, fantasy sports leagues, and fishing derbies.

Calcutta pools operate much like pari-mutuel betting, in that all the money wagered on a sporting event is pooled together. In a Calcutta pool, an auction is held before a sporting event, and bettors bid for the opportunity to bet on a particular player. For example, before a golf tournament the pool participants bid against each other for the right to bet on a particular golfer. The money collected during the auction becomes the wagering pool. It is divided among the "owners" of the best finishing players and the pool sponsor. Calcutta pools are most often associated with rodeos and golf tournaments.

A sports tab game is one in which players purchase a numbered tab from a game card containing 100 tabs with different number combinations. Bettors win money or prizes if their numbers match those that are associated with a sporting event—for example, digits in the winning team's final score. The cost of a sports tab is limited by law to less than $5. Operators of sports tab games (except charities) are allowed to take no more than 10% of the

TABLE 8.3

Amount wagered on Nevada sports betting, by sport, 1992–2016

[in thousands of dollars]

	Football	Basketball	Baseball	Parlay	Other	Total
1992	735,799	509,608	433,333	57,584	64,434	1,800,783
1993	864,310	549,873	455,874	66,717	72,093	2,006,283
1994	1,016,112	604,702	373,000	67,800	74,317	2,136,998
1995	1,039,936	679,704	533,023	71,088	96,031	2,428,593
1996	999,964	680,739	608,373	78,324	114,385	2,483,312
1997	949,086	637,477	566,172	66,882	132,166	2,431,382
1998	915,937	557,723	600,513	68,280	134,433	2,269,062
1999	980,276	610,333	678,254	63,510	129,433	2,470,407
2000	914,212	638,993	567,537	65,301	137,822	2,323,377
2001	806,319	567,179	495,657	58,004	112,780	2,042,855
2002	860,000	513,503	386,783	66,431	109,444	1,936,544
2003	826,244	489,167	378,875	63,984	106,634	1,863,678
2004	969,129	543,661	426,479	59,943	86,813	2,087,273
2005	1,051,198	579,396	464,067	62,864	100,176	2,257,174
2006	1,136,397	635,365	457,541	68,265	131,147	2,427,605
2007	1,176,512	687,193	528,792	68,616	134,000	2,594,191
2008	1,122,229	738,268	522,013	64,614	131,486	2,579,225
2009	1,093,596	803,529	488,630	58,981	126,848	2,568,362
2010	1,185,462	830,422	528,668	56,707	158,563	2,762,267
2011	1,341,788	737,432	558,011	62,633	179,465	2,877,935
2012	1,566,499	975,153	693,687	58,178	157,692	3,449,533
2013	1,622,791	1,053,132	681,007	57,658	204,889	3,622,107
2014	1,749,723	1,109,182	721,831	57,782	266,990	3,901,117
2015	1,698,477	1,251,732	897,312	54,070	333,174	4,237,422
2016	1,692,820	1,402,779	1,045,584	60,255	307,216	4,509,753

Note: This is the total amount bet in each sport.

SOURCE: David G. Schwartz, "Drop by Sport, 1992–2016," in *Nevada Sports Betting Totals: 1984–2016*, University of Nevada, Las Vegas,Center for Gaming Research, 2017, http://gaming.unlv.edu/reports/NV_sportsbetting.pdf (accessed June 30, 2017)

TABLE 8.4

Nevada sports betting revenues, by sport, 1989–2016

[in thousands of dollars]

	Football	Basketball	Baseball	Parlay	Other	Total
1989	19,042	6,485	6,616	9,865	3,769	45,776
1990	16,905	6,611	10,281	11,694	3,339	48,839
1991	17,584	15,827	5,000	10,455	3,432	52,300
1992	19,793	10,396	7,800	9,939	2,674	50,602
1993	30,683	17,266	9,391	17,073	620	75,035
1994	63,507	19,290	10,071	25,554	4,028	122,450
1995	32,550	18,352	4,584	19,855	3,774	79,415
1996	27,599	19,333	10,099	21,594	−2,139	76,486
1997	32,174	14,152	14,494	19,924	6,833	89,718
1998	28,852	21,305	4,684	15,841	6,399	77,375
1999	49,700	27,526	4,273	19,739	7,300	109,192
2000	49,276	27,285	19,353	20,942	6,643	123,836
2001	46,444	29,550	17,348	18,689	6,045	118,077
2002	40,334	30,348	11,062	21,796	6,884	110,383
2003	48,831	34,046	12,124	21,070	6,558	122,630
2004	46,712	30,445	9,084	19,943	6,320	112,504
2005	40,366	38,356	26,127	15,628	5,700	126,176
2006	91,139	46,191	22,145	21,087	10,977	191,538
2007	73,532	37,452	25,382	20,283	11,792	168,363
2008	39,278	43,484	23,595	20,263	9,822	136,441
2009	48,665	38,248	21,402	19,210	8,854	136,380
2010	56,428	39,362	22,627	18,997	13,684	151,096
2011	44,279	48,818	19,642	17,600	10,391	140,731
2012	68,456	47,880	30,106	13,573	10,045	170,062
2013	80,815	59,186	29,079	19,967	13,789	202,838
2014	113,732	54,239	21,294	21,281	16,500	227,045
2015	82,546	72,976	39,392	15,983	20,890	231,787
2016	91,243	66,632	32,204	10,412	18,648	219,174

SOURCE: Adapted from David G. Schwartz, "Sports Betting Win by Sport (Total Amount), 1989–2016," in *Nevada Sports Betting Totals: 1984–2016*, University of Nevada, Las Vegas, Center for Gaming Research, 2017, http://gaming.unlv.edu/reports/NV_sportsbetting.pdf (accessed June 30, 2017)

total amount wagered to cover their expenses (charities are allowed to take 50%). Sports tab sellers must obtain a license from the state and pay licensing fees and gaming taxes.

OTHER STATES. Other states that offer limited sports gambling are Washington, which permits $1 bets on race cars, and New Mexico, where small bets on bicycle races are legal. Office pools on sporting events are legal in a few states as long as the operator does not take a commission.

Fantasy Sports

During the first decade of the 21st century a new type of legal sports betting emerged: fantasy sports. In fantasy sports players choose real professional players for fictional teams; the teams compete against one another based on real-world performance. The Unlawful Internet Gambling Enforcement Act of 2006, which made bank transactions to online gaming websites illegal, specifically exempted fantasy sports games, and hundreds of online websites have since been created to allow fantasy sports players to wager on their activities. In 2007 the U.S. Court of Appeals for the Eighth Circuit ruled that such companies do not need to pay license fees to professional sports leagues because the First Amendment guarantees their right to use players' names and statistics. In "Industry Demographics" (2017, http://fsta.org/research/industry-demographics/), the Fantasy Sports Trade Association indicates that in 2017 an estimated 59.3 million people in the United States and Canada were involved in a fantasy sports league.

Joshua Brustein reports in "Fantasy Sports and Gambling: Line Is Blurred" (NYTimes.com, March 11, 2013) on the growing concerns regarding the lack of regulation of fantasy sports betting. One of the main reasons fantasy sports betting has not fallen under legal definitions of gambling is that the process of assembling a successful fantasy team to compete over the course of an entire season more closely resembles a game of skill, for which players can be rewarded, than a game of chance. However, as websites offering fantasy sports wagering have proliferated, they have introduced newer forms of fantasy betting that closely resemble other forms of gambling. Brustein explains, "Sometimes, they operate like season-long fantasy sports, in which players enter teams and compete against one another. Other variations include 'pick five,' in which players are presented with pairs of players, then wager on who will perform better; and player exchanges, in which users buy and sell stock in athletes." Some states have laws that restrict the kinds of wagers residents can make on fantasy sports, but in other states the activity is seen by many as a form of unregulated gambling. Furthermore, there are signs that the

TABLE 8.5

Nevada sports betting revenue percentages, by sport, 1989–2016

	Football	Basketball	Baseball	Parlay	Other
1989	41.60%	14.17%	14.45%	21.55%	8.23%
1990	34.61%	13.54%	21.05%	23.94%	6.84%
1991	33.62%	30.26%	9.56%	19.99%	6.56%
1992	39.12%	20.54%	15.41%	19.64%	5.28%
1993	40.89%	23.01%	12.52%	22.75%	0.83%
1994	51.86%	15.75%	8.22%	20.87%	3.29%
1995	40.99%	23.11%	5.77%	25.00%	4.75%
1996	36.08%	25.28%	13.20%	28.23%	−2.80%
1997	35.86%	15.77%	16.16%	22.21%	7.62%
1998	37.29%	27.53%	6.05%	20.47%	8.27%
1999	45.52%	25.21%	3.91%	18.08%	6.69%
2000	39.79%	22.03%	15.63%	16.91%	5.36%
2001	39.33%	25.03%	14.69%	15.83%	5.12%
2002	36.54%	27.49%	10.02%	19.75%	6.24%
2003	39.82%	27.76%	9.89%	17.18%	5.35%
2004	41.52%	27.06%	8.07%	17.73%	5.62%
2005	31.99%	30.40%	20.71%	12.39%	4.52%
2006	47.58%	24.12%	11.56%	11.01%	5.73%
2007	43.67%	22.24%	15.08%	12.05%	7.00%
2008	28.79%	31.87%	17.29%	14.85%	7.20%
2009	35.68%	28.05%	15.69%	14.09%	6.49%
2010	37.35%	26.05%	14.98%	12.57%	9.06%
2011	31.46%	34.69%	13.96%	12.51%	7.38%
2012	40.25%	28.15%	17.70%	7.98%	5.91%
2013	39.84%	29.18%	14.34%	9.84%	6.80%
2014	50.09%	23.89%	9.38%	9.37%	7.27%
2015	35.61%	31.48%	16.99%	6.90%	9.01%
2016	41.63%	30.40%	14.69%	4.75%	8.51%

Note: This is the percent of total sports betting win represented by each category.

SOURCE: David G. Schwartz, "Sports Betting Win By Sport (Percentage), 1989–2016," in *Nevada Sports Betting Totals: 1984–2016*, University of Nevada, Las Vegas, Center for Gaming Research, 2017, http://gaming.unlv.edu/reports/NV_sportsbetting.pdf (accessed June 30, 2017)

major professional sports leagues are also beginning to see the newer types of betting in this way.

Operators of fantasy sports gaming faced a major legal challenge in 2013, when FanDuel, a leading operator of fantasy sports daily contests, was sued in an Illinois federal district court by a player who lost money at the company's website. In "FanDuel Secures an Important Victory in Daily Fantasy Sports Lawsuit, However Plaintiff Plans to Appeal" (Forbes.com, October 9, 2013), Marc Edelman of Baruch College notes that the plaintiff, Christopher Langone, argued that FanDuel's operations amounted to illegal games of chance, and thus he was entitled to recovery of his losses under the terms of the Illinois Loss Recovery Act. The court dismissed the lawsuit in October 2013 on a number of grounds, including the plaintiff's failure to establish that his losses met the minimum necessary threshold to fall under the relevant statute. The court did not, however, decide the crucial issue of whether daily fantasy sports contests (i.e., fantasy sports wagers made on a daily basis rather than on a season-long basis) constituted legal games of skill or illegal games of chance.

Nevertheless, by mid-decade some states started viewing fantasy sports betting as a form of gambling. In October 2015 the Nevada Gaming Control Board began requiring sites such as FanDuel and DraftKings to obtain a gaming license to operate within the state. The following month Eric T. Schneiderman (1954–), the attorney general of New York, banned the two sites from operating in New York. The prohibition remained in place until August 2016, when the New York governor Andrew Cuomo (1957–) signed legislation that officially classified daily fantasy sports as a game of skill, rather than chance.

ILLEGAL SPORTS GAMBLING

Estimates on the amount of money that is spent on illegal sports betting vary widely. Hobson suggests that gamblers wager anywhere between $80 billion and $380 billion on illegal sports betting each year. The American Gaming Association (AGA) reports in *Law Enforcement Summit on Illegal Sports Betting: After-Action Report* (September 20, 2016, https://www.americangaming.org/sites/default/files/After%20Action%20Report_PDF-Web.pdf) that more recent studies indicate that illegal sports betting accounts for roughly $150 billion in illicit wagers annually. In other words, the $4.5 billion spent on legal sports betting in 2016 represented only 3% of the total amount wagered on sporting events that year. The AGA further notes that of all the money wagered on Super Bowl 50 and the NCAA men's basketball tournament in 2016, approximately 97% came through illegal betting. According to the AGA, the $150 billion spent each year on illegal sports wagers is roughly double the value of revenues reported by Google in 2015, and exceeds the annual revenues of 442 Fortune 500 companies.

In "How Much Sports Gambling Is Going On out There?" (ESPN.com, July 10, 2012), Peter Keating suggests that once illegal and casual wagering are taken into account, sports betting may be one of the most popular forms of gambling in the United States. In 2012 ESPN commissioned a survey of 1,148 U.S. residents aged 16 years and older to examine the subject of legal and illegal sports betting. The survey found that among the 50% of Americans who had placed a sports bet in the 12 months preceding the survey, 78% had either equaled or exceeded the amount that they had bet in the previous year. The survey also found that sports betting was, for most participants, fully integrated into their experience of being fans: 59% reported betting based on "gut feelings" and opinions, 66% claimed they would prefer to see their favorite team win a championship than to win a large amount of money by betting against the team, and only 15% considered their sports betting to be primarily motivated by the desire to make money.

Keating further notes that the ESPN survey revealed a growing comfort, among younger people, with the idea of sports betting, in part because of the way "new technology is constantly making private pursuits of pleasure more convenient and acceptable." Among survey respondents, those aged 21 to 29 years were five times more likely to have engaged in online sports betting than those aged 60 years and older. Among the 21- to 29-year-old respondents, 41% agreed that online sports gambling is "perfectly harmless," and nearly two-thirds agreed with the idea that sports betting is "no different," in moral terms, than buying a lottery ticket.

However, according to Keating, the ESPN survey found that a majority of Americans distrust the effect that gambling has on sports: 52% of all respondents expressed the belief that some sporting events are fixed and 60% of those who bet regularly on sports felt that some contests are fixed. Perhaps as a result, only 37% of the survey sample expressed support for legalizing sports gambling nationwide. In demographic terms, sports betters were comparatively wealthy, and the overwhelming majority of them were men. "The typical sports bettor is a 42-year-old married guy living in the suburbs, with a household income of $74,000 a year," Keating explains. "Among heavy bettors, 6% have an annual income of $150,000 or more, twice the number of nonbettors in the general population who make that kind of coin."

Most illegal bets on sporting events are placed with bookies, although illegal Internet gambling is increasing in popularity. The most popular forms of sports betting involve betting on the teams of the major professional and college sports leagues, but other illegal "sporting" events are also popularly associated with gambling, such as cockfighting and dogfighting.

The Link to the Nevada Sports Books

Most illegal sports books use the odds posted by the Nevada sports books because these are well publicized. Nevada sports books also provide illegal bookies with a means for spreading the risk on bets: illegal bookies who get a lot of action on one side of a bet often bet the other side with the Nevada sports books to even out the betting.

Transmitting gambling information across state lines for the purpose of placing or taking bets is illegal. News items about point spreads (the predicted scoring difference between two opponents) can be reported for informational and entertainment purposes only, but betting lines are still published by many U.S. newspapers. The News Media Alliance (formerly the Newspaper Association of America until 2016), which represented approximately 2,000 print and digital newspapers in 2017, defends the practice as free speech that is protected under the First Amendment of the U.S. Constitution. The alliance claims that readers want to see the lines for informational purposes (to learn which teams are favored to win) and not necessarily for betting purposes. Although the NCAA argues that a ban on all college sports wagering would pressure newspapers to stop publishing point spreads, the AGA and others counter that betting lines would still be accessible through independent sports analysts, offshore Internet gambling sites, and other outlets.

The Link to Organized Crime

Illegal sports gambling has long been associated with organized crime in the United States. During the 1920s and early 1930s mobsters set up organized bookmaking systems across the country, including two illegal wire services, Continental Wire Service and Trans America Wire, which operated under the direction of the gangster Al Capone (1899–1947). The legal wire service, Western Union, was prohibited by law from transmitting race results until the races were officially declared over. Sometimes this declaration was delayed for several minutes following the completion of the races, so mobsters reported the winners on the illegal wire services to prevent bettors from taking advantage of these delays by posting winning bets before the official results were wired.

During the 1950s the federal government cracked down on organized crime and eventually drove mobsters out of the Nevada casino industry. As the casinos were taken over by corporations, organized crime strengthened its hold on illegal bookmaking. Although law enforcement officials acknowledge that many "independent" bookies operate throughout the country, the big money in illegal sports gambling is still controlled by organized crime.

ANIMAL FIGHTING

Gambling on animal fights has a long history in the United States. Most staged animal fights involve cocks (roosters) or dogs that are specially bred and trained. Such fighting is usually associated with rural areas, but urban police reports about cockfighting and dogfighting have increased since the early 1990s, mostly because the contests have become popular among street gangs. Animal fights are of particular concern to law enforcement authorities because large amounts of cash and weapons are usually present.

Although reliable national statistics on the prevalence of animal fighting are elusive, the website Pet-Abuse.com maintains the Database of Criminal Animal Cruelty Cases (http://www.pet-abuse.com/pages/cruelty_database.php), which lists media reports of animal fighting and other forms of animal cruelty. As of August 2017, the database recorded allegations against eight offenders (two in California and one each in Florida, Georgia, Minnesota, New Jersey, South Carolina, and Tennessee), all of which happened in 2014, the most recent year for which data were available. Some of the cases involved the discovery of a single dog or a small number of roosters determined to have died from fighting, whereas others involved the seizure of several animals. For example, an April 22, 2014, raid in Bordeaux,

Tennessee, led to the rescue of 36 dogs and the arrest of their owner; and an April 19, 2014, raid in McBee, South Carolina, led to the seizure of 132 birds and the charging of 46 people who were arrested for participating in cockfighting.

Cockfighting

Cocks fight and peck one another naturally to establish a hierarchy within their social order. These altercations rarely lead to serious injury. By contrast, fighting cocks are specially bred and trained by humans to be extremely aggressive. They are given stimulants, steroids, and other drugs to heighten their fighting nature. Before a fight, the cocks have sharp spikes, called gaffs, attached to their legs, and then they are thrown into a pit. They slash and peck at one another, often until death. Spectators wager on the outcome of these fights.

In "Cockfighting Fact Sheet" (2017, http://www.humanesociety.org/issues/cockfighting/facts/cockfighting_fact_sheet.html?credit=web_id80597821), the Humane Society of the United States indicates that cockfighting was illegal in all 50 states in 2017. It also notes that cockfighting was a felony in 39 states. The federal Animal Welfare Act prohibits the interstate transport of birds for cockfighting.

Dogfighting

Spectators huddle around pits or small, boarded arenas to watch dogfights. They place bets on the outcome of the contests, which can go on for hours, sometimes to the death. Dogs are specially bred and trained for such fights—the American pit bull terrier is the most common breed because of its powerful jaws. Authorities report that the dogs are often draped in heavy chains to build muscle mass and systematically deprived of food and water. Stolen and stray pet dogs and cats are commonly used as bait to train the fighters. The smaller animals are stabbed or sliced open and thrown to the fighting dogs to enhance their blood lust. The dogs are often drugged to increase their aggressiveness. Dogfighting is a felony in all 50 states.

One of the highest profile cases of dogfighting involved Michael Vick (1980–), a professional football player who was indicted on charges linked to dogfighting in July 2007, at which time he was a quarterback for the Atlanta Falcons. His alleged offenses included federal conspiracy charges involving crossing state lines to participate in illegal activity, as well as buying and sponsoring pit bulls in fighting ventures. Vick and his associates were charged with having bought property in Virginia on which to stage dogfights over a six-year period. Dogs who performed poorly were killed violently by the dogfighting ring. Vick himself admitted to killing several dogs. He pleaded guilty and was sentenced to 23 months in prison. Following his release from prison, Vick joined the Philadelphia Eagles at the beginning of the 2009 NFL season.

THE EFFECTS OF ILLEGAL SPORTS GAMBLING ON SOCIETY

Money and Crime

Because the vast majority of sports gambling that occurs in the United States is illegal, it is difficult to determine its economic effects. Although some particularly skilled handicappers are able to make six-figure incomes as full-time sports gamblers, the most prominent beneficiaries of illegal sports gambling are the bookmakers. Large bookmaking operations that are overseen by organized crime groups take in billions of dollars each year. The betting stakes are high and the consequences for nonpayment can be violent. Small independent bookies typically operate as entrepreneurs, taking bets only from local people they know well. Illegal bookmaking cases reported in the media range from multimillion-dollar enterprises to small operations run by one person.

Sports Tampering

The Federal Bureau of Investigation (FBI) defines sports tampering in *National Incident-Based Reporting System—Volume 1: Data Collection Guidelines* (August 2000, https://ucr.fbi.gov/nibrs/nibrs_dcguide.pdf) as "to unlawfully alter, meddle in, or otherwise interfere with a sporting contest or event for the purpose of gaining a gambling advantage." The most common form is point-shaving, which occurs when a player deliberately limits the number of points scored by his or her team in exchange for payment of some sort. For example, if a basketball player purposely misses a free throw in exchange for a fee, that player is participating in a point-shaving scheme.

Gambling has led to some famous sports scandals, mostly in college basketball games. Nevertheless, any link between an athlete and gambling gives rise to suspicions about the integrity of the games in which that athlete participates.

The professional baseball player Pete Rose (1941–) is an example. On September 11, 1985, at Riverfront Stadium in Cincinnati, Ohio, Rose broke Ty Cobb's (1886–1961) all-time hit record. Before the end of the decade, however, Rose was under investigation by the Major League Baseball commissioner and by federal prosecutors for betting on sporting events and associating with known bookies. He agreed to leave baseball, and the case was dropped. At the time, Rose denied ever betting on baseball games. However, in January 2004 he admitted that he had bet on baseball games while he managed the Cincinnati Reds during the late 1980s.

THE INTEGRITY OF COLLEGE SPORTS. The popularity of sports gambling among college students makes the NCAA a particular locus of concern regarding the effects of gambling. In "Gambling on College Sports: What's the Big Deal?" (2017, http://www.fiusports.com/documents/2014/6/17/NCAA_gambling_1.pdf?id=568), the association notes that, according to FBI estimates, more than $2.5 billion in illegal bets are placed annually on the March Madness college basketball tournament, more than is typically wagered on the Super Bowl. The NCAA further observes that sports wagering is the typical entry point for young people who become gamblers as adults and that as a result of social media and other technological advances, college athletes and others affiliated with collegiate sports teams are under increasing demands to supply gamblers with inside information that might influence sports betting.

According to the NCAA, in *NCAA Student-Athlete Gambling Behaviors and Attitudes: 2004–2012* (May 2013, http://fs.ncaa.org/Docs/public/pdf/ncaa_wagering_prelim_may2013.pdf), the most recent study on this topic as of August 2017, 25.7% of male student athletes gambled on sports in 2012, down from 29.5% in 2008 and up from 23.5% in 2004. The proportion of female student athletes who gambled on sports was much lower, at 5.2%. Rates of probable pathological gambling were low among male student athletes, at 0.7% in 2012, and even lower among female student athletes, at less than 0.1%.

The NCAA opposes both legal and illegal sports gambling in the United States. Bylaw 10.3 of the NCAA prohibits staff members and student athletes from engaging in gambling activities that are related to college and professional sporting events. It also forbids them from providing any information about collegiate sports events to people involved in organized gambling activities.

The NCAA opposes sports gambling for the following reasons:

- It attracts organized crime.
- The profits fund other illegal activities, such as drug sales and loan-sharking (lending money at an excessive rate of interest).
- Student athletes who are involved can become indebted to bookies, leading to point-shaving schemes.

CHAPTER 9
INTERNET GAMBLING

Internet gambling represents a new frontier for the gaming industry, for government regulators, and for the residents of the states and countries that allow it. The Internet has transformed numerous global industries almost beyond recognition, both through e-commerce (the selling of goods in virtual rather than in physical stores) and by offering radical new distribution platforms for less tangible goods and services. For example, the music and news industries have seen their traditional business models almost entirely upended by the Internet, which allows anyone to distribute music and news online very inexpensively.

The gambling industry, which ultimately sells the experience of gambling rather than physical products, is one that is uniquely susceptible to being revolutionized by the Internet. Such radical change creates enormous possibilities for new entrants to the marketplace, as well as the potential for a wide range of unintended positive and negative consequences. However, as of August 2017, legal hurdles to the mass adoption of online gambling in the United States had only recently been lifted, and forms of Internet gambling were legal in only three states: Delaware, Nevada, and New Jersey. Accordingly, the U.S. gaming industry was only beginning to respond to the promises and threats represented by legalized online gambling.

THE HISTORY AND CHANGING LEGAL STATUS OF INTERNET GAMBLING

Between the mid-1990s and the first half of the first decade of the 21st century Internet gambling in the United States occurred in the absence of significant regulation or legal restrictions. The Spectrum Gaming Group notes in *Gambling Impact Study* (October 28, 2013, http://www.leg.state.fl.us/GamingStudy/docs/FGIS_Spectrum_28Oct2013.pdf) that online poker enjoyed increasing popularity during that time. This was a function not only of the rapidly increasing role that the Internet played in the lives of most Americans but also of the increasing popularity of the World Series of Poker (WSOP), which was founded in 1970 and which became a prominent fixture on television in the decades that followed.

According to the Spectrum Gaming Group, by 2000 the Internet had begun to transform the WSOP, allowing hundreds of players from around the world to play introductory rounds online to qualify for the live televised event. The early WSOP tournaments involved only 12 players, and by the 1980s the tournament had grown to around 50 competitors. In 2000, thanks to the Internet, 450 players vied for a place in the final tournament field, and in 2003 Chris Moneymaker, an amateur who had qualified with an initial $40 wager on the website PokerStars, outplayed 838 other people on his way to winning $2.5 million. Moneymaker's high-profile wins led to an explosion in the WSOP field the following year, as 2,576 players sought to qualify for the tournament. In 2006 the size of the WSOP field peaked at 8,773 contestants.

The increasing popularity of Internet poker did not, however, go unnoticed by the U.S. gaming industry or by government officials. Internet gambling to that date existed in a legal limbo that kept major casino operators from participating, and most in the tribal and commercial casino industries were opposed to legalization. The Internet gambling sector was highly fragmented and replete with small operators, most of them located outside of the United States. In most cases, these businesses paid no taxes in the United States, and many paid only a minimal amount in their home country. In *An Analysis of Internet Gambling and Its Policy Implications* (May 31, 2006, https://www.americangaming.org/sites/default/files/research_files/wpaper_internet_0531.pdf), David O. Stewart of Ropes and Gray, LLP, of Boston, Massachusetts, indicates that by 2005 two-thirds of Internet gambling operations were located in small Caribbean and Central American countries that provided little or no government oversight of the industry. In March 2005 the island nation of Antigua and Barbuda in the Caribbean served as the headquarters for 536 gambling sites, the most of any country. The sites were only required to pay 3% of their gambling revenues (winnings after payout to customers) to the government with a ceiling of $50,000 per month.

The U.S. government intervened in the online gambling market in 2006 with congressional passage of the Unlawful Internet Gambling Enforcement Act (UIGEA), which prohibited Internet gambling, although it provided a specific exclusion for fantasy sports and did not address the legality of online lottery purchases or betting on horse races. Under the UIGEA banks and credit card companies were prohibited from transferring Americans' money to Internet gambling sites. Many casual gamblers turned away from online gambling sites when the law was enacted, and although serious Internet gamblers continued to find ways of transferring funds to online casinos and card rooms, the initial heyday of online poker in the United States was over.

By late 2006 many of the larger, publicly traded Internet gambling companies, such as PartyPoker, had stopped accepting American customers altogether to avoid any conflicts with the U.S. government. However, other online gambling outlets continued to allow U.S. residents to place bets on their sites in violation of the new law. The U.S. Department of Justice put an end to this state of affairs in April 2011, when it indicted three online poker operators—Full Tilt Poker, Absolute Poker, and PokerStars—and confiscated both their assets and the domain names under which they operated. Those serious U.S. poker players who had persisted in playing online following the passage of the UIGEA largely ceased doing so, and Europe and Asia became the primary markets for Internet gambling operators. These events are reflected in the revenue levels for the North American Internet poker industry, as published by the Spectrum Gaming Group. North American online poker revenues plummeted from more than 1 billion euros in 2010 to under 500 million euros in 2011, even as the European Internet poker market continued to see steady growth during this period.

A reversal came in late 2011, when the landscape for legal online gambling in the United States changed dramatically. This occurred as a result of an opinion issued by the Department of Justice in response to the state lotteries of New York and Illinois, which had asked for clarification of federal policy regarding online lottery purchases. The Department of Justice's public opinion, *Whether Proposals by Illinois and New York to Use the Internet and Out-of-State Transaction Processors to Sell Lottery Tickets to In-State Adults Violate the Wire Act* (September 20, 2011, https://www.justice.gov/sites/default/files/olc/opinions/2011/09/31/state-lotteries-opinion_0.pdf), stated that the federal Wire Act of 1961 did not prohibit online lottery sales or other forms of gambling other than sports betting. Although this opinion did not eliminate all doubt about the legality of Internet gambling, it opened the door to state passage of laws allowing for forms of Internet gambling other than sports betting. Delaware, Nevada, and New Jersey quickly moved to take advantage of this opening. Delaware officially legalized online gambling in July 2012, and Nevada and New Jersey both followed in February 2013. All three states implemented their new online gambling laws, which differed significantly in their details, in 2013. According to the Spectrum Gaming Group, in the years that followed North American Internet poker revenues saw a considerable increase, topping 1.7 billion euros by 2015.

Delaware's law established the state lottery as the only legal Internet gambling operator in the state, and it authorized the lottery to offer lottery ticket sales as well as selected casino games. Nevada's law authorized casinos and their business partners to operate Internet gambling sites, but only poker was allowed. New Jersey authorized Atlantic City casinos to operate Internet gambling sites, and the casinos were allowed to offer all games that were available in their brick-and-mortar facilities.

Industry analysts widely expect that other states will follow the lead of Delaware, Nevada, and New Jersey in the years to follow. Dustin Gouker notes in "Online Gambling and Poker Bill Tracker" (OnlinePokerReport.com, March 1, 2017) that in 2017 bills legalizing various forms of online gambling were under consideration in eight states: California, Illinois, Massachusetts, Michigan, New York, New Hampshire, Pennsylvania, and West Virginia.

Those opposed to the expansion of Internet gambling did not, however, cease fighting in the wake of the Department of Justice's 2011 opinion. In "Concerns Growing over How Spread of Online Gambling Will Play Out" (LATimes.com, February 16, 2014), Michael Hiltzik indicates that even as online gambling implementation began in Delaware, Nevada, and New Jersey, Sheldon Adelson (1933–), the chair and chief executive officer of Las Vegas Sands, was backing the Coalition to Stop Internet Gambling, which aired television commercials suggesting that Internet gambling would lead to unprecedented moral decay and social problems. Adelson was also a major political donor whose many sizeable campaign contributions, according to Steve Tetreault, in "New Bill Would Prohibit Internet Gambling, Including Where Already Legal" (ReviewJournal.com, March 26, 2014), included $15,600 to the U.S. senator Lindsey Graham (1955–; R-SC). Graham was one of the two lead sponsors of a 2014 U.S. Senate bill that would make all Internet gambling illegal in the United States.

By 2017 the issue of Internet gambling divided the casino industry. Whereas Adelson and the Las Vegas

Sands were vocal opponents of legalization, a number of competitors, including MGM Resorts International and the Caesars Entertainment Corporation, backed the expansion of legal online gambling. Kate O'Keeffe reports in "Online Gambling Suffers Setback" (WSJ.com, May 21, 2014) that because of the split within the gaming industry, the American Gaming Association (AGA) withdrew its support for the further legalization of Internet gambling in May 2014, officially declaring that it would take no position on the matter.

ONLINE GAMES

Online casinos offer many of the same games that are available in land-based casinos, such as poker, blackjack, roulette, slot machines, and bingo. Bet denominations range from pennies to thousands of dollars. Poker websites have card rooms where players compete against each other rather than against the house. This is an example of person-to-person betting. To make a profit on these sites, the casino operators take a small percentage of the winning hand. According to the research firm MarketLine, which profiles the Internet gambling industry in the 2016 edition of *Global Online Gambling* (February 2016, https://store.marketline.com/report/ohme5418--global-online-gambling/), between 2011 and 2015 online gambling revenues grew an average of 5.4% annually and were estimated to top $41.4 billion in 2015.

Online casino games operate in much the same way as the electronic games found in actual casinos. Both depend on random number generators: slot machines have a computer chip built in, and online games have random number generators written into their programming. Slot machine payoff percentages at actual casinos are dictated by the state in which they are located, whereas online payoffs are not. However, online providers who never have winners would not have return customers, so their programs are designed to pay out a particular percentage. Online games are particularly appealing to people who enjoy card games because the betting limits are much lower than they are in actual casinos. For example, an online gambler can play blackjack for $1 per hand, whereas many land-based casinos set a $10- or $25-per-hand minimum.

A handful of U.S. state lotteries had already begun offering online subscription services prior to the Department of Justice's 2011 opinion. Following the department's clarification, the Illinois Lottery and the Georgia Lottery began offering online sales, and other lotteries were expected to follow suit as a means of countering slowed growth. Delaware began offering online lottery tickets in 2013 as part of its entry into legalized online gambling, and Minnesota became the first state to offer scratch-off tickets for sale online in 2014.

The Spectrum Gaming Group notes in *Gambling Impact Study* that another form of online wagering has the potential to broaden the very definition of gambling and the potential size of the online market. Players and website operators have already monetized a number of Internet-based games that, although they are not widely associated with casinos and the gaming industry, could fall under the legal definition of gambling as the states move to regulate online wagering. For example, as is discussed in Chapter 8, daily contests based on fantasy sports leagues represent a burgeoning market that complicates fantasy sports operators' categorization as non-gambling enterprises, and this subset of fantasy sports play is expected to grow much larger if its legal status is clarified. Games played on social networking sites such as Facebook, including a number of casino-style games, represent another source of gambling revenue with virtually unlimited growth potential, given that these sites attract billions of people worldwide. According to the Spectrum Gaming Group, the five most popular casino games played on Facebook attracted 11.2 million daily participants in 2012.

However, the most popular form of Internet gambling worldwide is sports betting. In the 2015 edition of *Global Online Gambling* (2015, https://store.marketline.com/report/ohip1084--global-online-gambling/), MarketLine notes that sports betting accounted for 45.8% of all global online gambling revenues in 2013. Online sports betting is the only form of Internet gambling expressly prohibited by federal law, according to the Department of Justice's 2011 opinion, but its prevalence in the United States is believed to be high. As is discussed in Chapter 8, some estimates suggest that illegal sports betting is one of the most prevalent forms of gambling in the United States. Although some online sports betting operators have withdrawn from the U.S. market since 2006, others continue to process bets by U.S. residents.

Meanwhile, by 2016 other forms of online gambling had begun to emerge. In "Here's How to Legally Gamble on the 2016 Race" (WashingtonPost.com, March 28, 2016), Jessica Contrera reports that during the 2016 presidential campaign prediction market websites such as PredictIt.org and PredictWise.com enabled online traders to place wagers on the outcome of primary races. Because these sites were technically compiling data that could prove valuable to researchers, Contreras notes, betting on prediction markets was excluded from the prohibition on political gambling in the United States.

REVENUE LEVELS AND PROJECTIONS

Global

Legislation in other countries is also expected to move in the direction of liberalization, driving continued growth in the online gambling industry. In the 2015 edition of *Global Online Gambling*, MarketLine projects that growth will maintain a steady pace between 2013 and 2018, during which the estimated average annual rate will be 9.6%. By 2018 the industry is expected to generate roughly $55.8 billion in total revenues.

United States

Although online gambling generates only a small fraction of overall casino revenues, Internet gaming showed signs of steady growth between 2013 and 2017. The Center for Gaming Research at the University of Nevada, Las Vegas, provides in *United States Online Gaming: Monthly Statewide and National* (February 2015, http://gaming.unlv.edu/reports/US_online_gaming.pdf) an early assessment of online gaming revenues in Nevada, New Jersey, and Delaware. Between April 2013 and November 2013 Nevada's online poker operations generated just over $5 million in revenues. This figure represented 5.9% of the $85.2 million in overall poker revenues reported for this span. Internet poker in Nevada showed modest growth the following year, when online poker games accounted for $8.1 million in revenues between February 2014 and November 2014, or 8.1% of the $100.5 million in total poker revenues generated during the same period. The Center for Gaming Research notes that the Nevada Gaming Control Board ceased providing separate financial data for Internet poker in December 2014.

In *Atlantic City Gaming Revenue: Annual Statistics for Total, Slot, Table, & Internet Win, 1978–2016* (January 2017, http://gaming.unlv.edu/reports/ac_hist.pdf), the Center for Gaming Research provides a detailed analysis of the growth of online gambling in New Jersey. In 2014, the first full year when Internet gaming was legal in New Jersey, the state's online gaming operations produced $122.9 million in revenues. In 2015 the total win generated by Internet gambling reached $148.9 million, an increase of 21.2% over the previous year. Meanwhile, overall revenues for the New Jersey casino industry fell 6.5%, from $2.7 billion in 2014 to $2.6 billion in 2015. Internet gaming in New Jersey continued to enjoy strong growth in 2016, when online gambling generated $196.7 million, or nearly 7.6% of the casino industry's total revenues of $2.6 billion for the year. Figure 9.1 provides a breakdown of monthly online gaming revenues for Atlantic City casinos in 2016.

According to the Center for Gaming Research, in *United States Online Gaming*, online gaming in Delaware includes poker, table games, and video lottery. In 2014, the first full year of Internet gambling in Delaware,

FIGURE 9.1

New Jersey online gambling revenues, by location, 2016

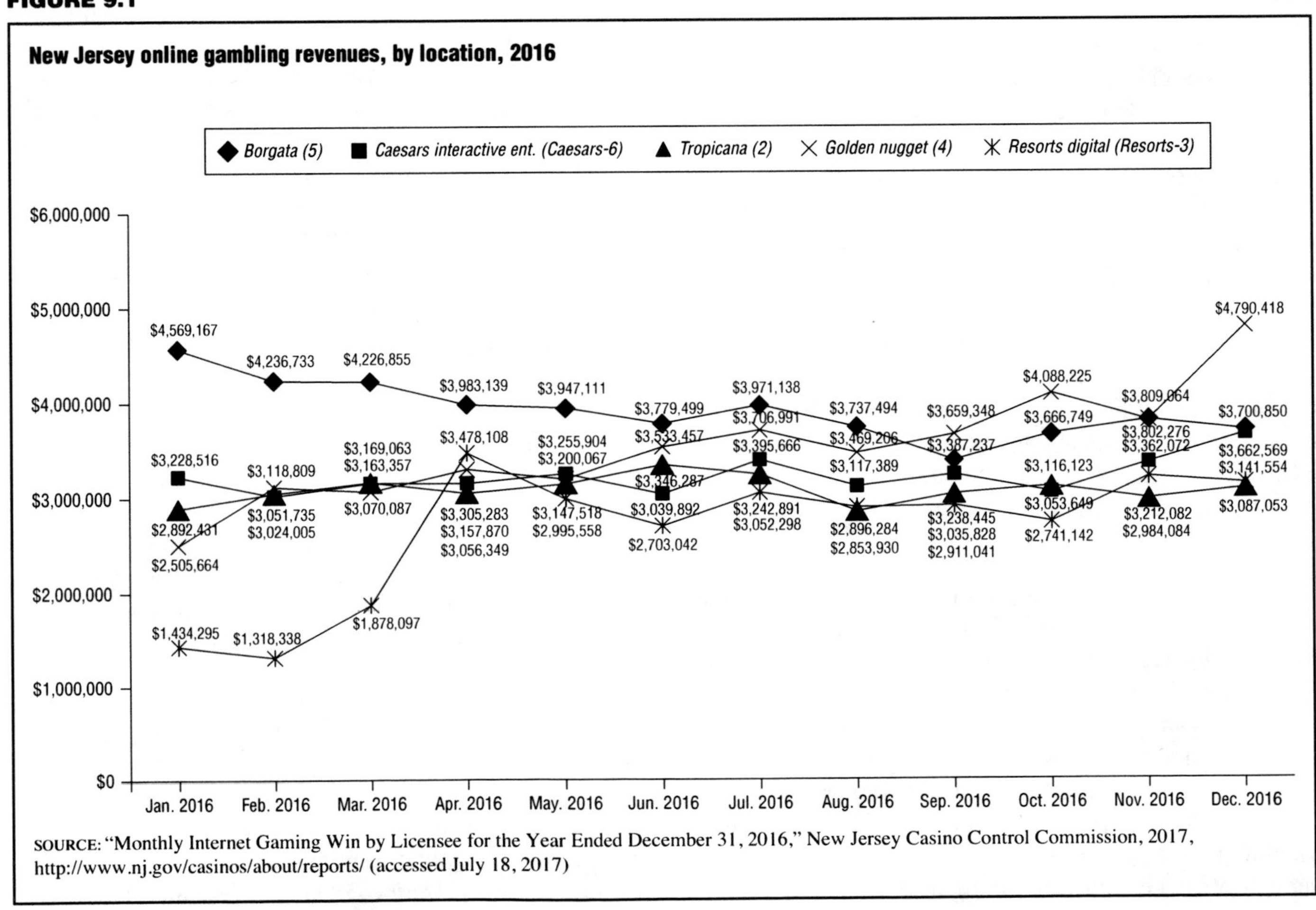

SOURCE: "Monthly Internet Gaming Win by Licensee for the Year Ended December 31, 2016," New Jersey Casino Control Commission, 2017, http://www.nj.gov/casinos/about/reports/ (accessed July 18, 2017)

TABLE 9.1

Delaware online gaming financial data, June 2016–May 2017

Year	Month	Game	Amount played	Amount won	Net earnings
2016	June	Table games	$4,925,915.89	$4,843,218.50	$82,697.39
		Video lottery	$3,531,515.62	$3,345,649.66	$185,865.96
		Poker rake & fee			$36,310.95
		Total	**$8,457,431.51**	**$8,188,868.16**	**$304,874.30**
	July	Table games	$3,075,538.09	$2,989,025.00	$86,513.09
		Video lottery	$3,075,538.10	$3,435,277.42	$160,878.27
		Poker rake & fee			$33,745.73
		Total	**$6,671,693.78**	**$6,424,302.42**	**$281,137.09**
	August	Table games	$3,569,682.10	$3,511,324.50	$58,357.60
		Video lottery	$2,826,908.90	$2,670,034.15	$156,874.75
		Poker rake & fee			$29,155.08
		Total	**$6,396,591.00**	**$6,181,358.65**	**$244,387.43**
	September	Table games	$5,085,096.57	$5,038,508.50	$46,588.07
		Video lottery	$2,900,116.78	$2,752,173.97	$147,942.81
		Poker rake & fee			$34,654.15
		Total	**$7,985,213.35**	**$7,790,682.47**	**$229,185.03**
	October	Table games	$3,105,425.41	$3,046,415.50	$59,009.91
		Video lottery	$3,520,904.96	$3,425,963.78	$94,941.18
		Poker rake & fee			$26,045.50
		Total	**$6,626,330.37**	**$6,472,379.28**	**$179,996.59**
	November	Table games	13,107,891.99	$13,082,473.00	$25,418.99
		Video lottery	$2,417,699.20	$2,259,928.52	$157,770.68
		Poker rake & fee			$23,056.88
		Total	**$15,525,591.19**	**$15,342,401.52**	**$206,246.55**
	December	Table games	$13,256,268.91	$13,174,062.50	$82,206.41
		Video lottery	$2,516,197.74	$2,394,328.20	$121,869.54
		Poker rake & fee			$25,911.15
		Total	**$15,772,466.65**	**$15,568,390.70**	**$229,987.10**
2017	January	Table games	$11,364,708.50	$11,318,628.00	$46,080.50
		Video lottery	$2,075,790.77	$1,952,675.62	$123,115.15
		Poker rake & fee			$25,474.33
		Total	**$13,440,499.27**	**$13,271,303.62**	**$194,669.98**
	February	Table games	$8,408,472.67	$8,294,223.00	$114,249.67
		Video lottery	$1,950,746.70	$1,833,789.01	$116,957.69
		Poker rake & fee			$19,367.78
		Total	**$10,359,219.37**	**$10,128,012.01**	**$250,575.14**
	March	Table games	$4,257,356.47	$4,221,727.00	$35,629.47
		Video lottery	$3,772,292.01	$3,673,825.40	$98,466.61
		Poker rake & fee			$17,715.38
		Total	**$8,029,648.48**	**$7,895,552.40**	**$151,811.46**
	April	Table games	$4,213,588.51	$4,185,607.00	$27,981.51
		Video lottery	$4,014,002.98	$3,858,732.68	$155,270.30
		Poker rake & fee			$19,280.01
		Total	**$8,227,591.49**	**$8,044,339.68**	**$202,531.82**
	May	Table games	$3,953,753.12	$3,872,317.00	$81,436.12
		Video lottery	$3,157,786.90	$2,957,794.84	$199,992.06
		Poker rake & fee			$20,693.45
		Total	**$7,111,540.02**	**$6,830,111.84**	**$302,121.63**

SOURCE: Adapted from "Delaware iGaming Net Proceeds (Unaudited)," in *iGaming*, Delaware Lottery, 2017, http://www.delottery.com/games/igaming/finalstats.asp and http://www.delottery.com/games/igaming/finalstats2016.asp (accessed June 30, 2017)

online gaming generated $2.1 million in revenues for the state. Table games accounted for $998,900, or 47.8% of all online gaming revenues generated that year. Over the next two and a half years, the state's online gambling operations experienced substantial growth. Between June 2016 and May 2017 Internet gambling in Delaware generated nearly $2.8 million in revenues. (See Table 9.1.) This figure represented an increase of 33% compared with revenues reported in 2014.

Obstacles and Growth Potential for Legal Online Operators

Although the laws allowing gambling in Delaware, Nevada, and New Jersey all differ significantly, they share a requirement mandating that players be physically present in the state where the game is being hosted. The online gaming platforms in each state use geolocation software (programs that can verify the geographical position of a computer or mobile device user) to verify that the player either lives in or is visiting the state. The systems further require registration verifying the player's age and other personal information. Accordingly, legal online gambling in the United States is unable to draw on the worldwide customer base that offshore Internet gambling operators enjoy.

The Spectrum Gaming Group notes in *Gambling Impact Study* that analysts expect states that legalize Internet gambling will enter into compacts similar to the cooperative state lottery compacts that paved the way for the multistate Powerball and Mega Millions games (see Chapter 7). If an increasing number of states legalize online gambling in the coming years, such multistate compacts will have the potential to broaden the appeal of online gambling by increasing the level of competition and liquidity (the ease with which financial transactions of any size can be accomplished without affecting the ability of the operator to continue offering games). This would be of particular value in online poker, where large numbers of participants are crucial. In such a scenario, even a state such as Nevada that offers only online poker could expect revenues much larger than those it saw during its initial introduction of Internet gambling.

In "Nevada, Delaware Ink Online Poker Pact to Expand Player, Revenue Pools" (LasVegasSun.com, February 25, 2014), Karoun Demirjian reports that the governors of Nevada and Delaware entered into the first such agreement in February 2014. Players would still be required to log in via their respective state's systems and to have their locations verified, but after doing so they would be able to engage in games involving players from both states. The casino operators would split the revenues from multistate games proportionately according to the individual players' locations. According to Demirjian, Jack Markell (1960–), the governor of Delaware, said at a news conference announcing the compact, "We hope [the deal] will serve as a model for multistate collaboration and that other states will see the benefits of the agreement and soon decide to join for themselves." Actual implementation of multistate play was subject to the decisions of the individual online gambling operators involved. As of August 2017, it was unclear when implementation might occur, but as Demirjian observes of the press conference announcing the compact, "It was clear that today's agreement was as much a blueprint for future interstate compacts as it was a contract in its own right."

Even with multistate compacts, the U.S. online gambling industry is likely to face continued competition from illegal operators. According to the article "Experts: Online Gambling Slowed by Illegal Action" (Associated Press, May 20, 2014), a panel of experts speaking at the East Coast Gaming Congress in Atlantic City in May 2014 noted that illegal offshore gambling companies continue to operate on a large scale and that this is a primary obstacle to the development of the legal industry in the United States. One of the panelists, Richard Schuetz, the gambling control commissioner of California, estimated that more than 1 million people in his home state alone used the Internet to gamble illegally and that illegal online gambling in the United States generated $300 million to $400 million in revenues per year. "Internet gambling exists in all 50 states today," David Rebuck, the director of the New Jersey Division of Gaming Enforcement, told the panel audience, "It's just not regulated."

Additionally, bringing Internet gambling under regulation has not necessarily assuaged some consumers' concerns. According to Rebuck, legal online gambling outlets require players to enter their Social Security number and credit card information to register, whereas unregulated sites allow people to play without divulging personal information, a fact that has likely driven many players with security concerns to illegal sites. Furthermore, a panelist affiliated with an offshore casino operator maintained that his company has conducted surveys demonstrating that only 10% of New Jersey residents realize that online gambling has been legalized in their state. Another panelist affiliated with an offshore company noted that credit card processing on gambling sites is woefully inefficient and too time consuming for the average online player.

ONLINE GAMBLERS

Given the rapidly shifting legal terrain of online gambling in the United States and the nascent state of the industry sector in those states where it was legal as of 2017, demographic profiles of Internet gamblers are currently provisional, at best. An increasing amount of research assessing the characteristics of Internet gamblers has been undertaken since the mid-1990s, both by analysts within the gaming industry and by scholars who study gambling, but much of it pertains to global online gambling prior to U.S. legalization.

In *Online Gambling Five Years after UIGEA* (May 18, 2011, https://www.americangaming.org/sites/default/files/research_files/final_online_gambling_white_paper_5-18-11.pdf), Stewart notes that, in spite of legal prohibitions, approximately 10 million Americans had gambled online during the previous decade. According to Stewart, between 2003 and 2010 Americans wagered roughly $30 billion on Internet gambling. This money was bet on a variety of games and contests, including sports betting, online poker, and electronic casino games such as blackjack, roulette, and slots. The AGA indicates in *2013 State of the States: The AGA Survey of Casino Entertainment* (2013, https://www.americangaming.org/sites/default/files/research_files/aga_sos2013_rev042014.pdf) that according to its survey of a representative pool of U.S. residents, 3% of adults placed some form of Internet wager in 2012.

In *Gambling Impact Study*, the Spectrum Gaming Group notes that one of the key advantages that online gambling offers the existing commercial casino industry is a diversification of its customer base. The group explains, "Both commercial and tribal casinos in the United States, as well as domestic lotteries find themselves in the same situation demographically. Their core player base is aging and not being fully replaced by a younger generation." Although surveys routinely show that nearly nine out of 10 U.S. residents use the Internet, those who use it most frequently are on average younger, wealthier, more educated, and more likely to be employed full time than the U.S. population as a whole. Assuming that the legalization of online gambling in the United States continues to involve existing casino operators, as is the case in Delaware, Nevada, and New Jersey, Internet games have the potential to broaden the industry's customer base and help it transition into a new generation.

According to Sally M. Gainsbury et al., in "How the Internet Is Changing Gambling: Findings from an Australian Prevalence Survey" (*Journal of Gambling Studies*, vol. 31, no. 1, 2015), surveys conducted globally between 2009 and 2012 suggest that, although Internet gamblers are typically also offline gamblers, they represent a distinct demographic subset of total gamblers. They are more likely than land-based gamblers to be male, young, better-educated, and employed full time, and their socioeconomic status is more likely to be higher than that of gamblers in general. Although most researchers agree that gamblers both online and offline are predominantly male, Abby McCormack, Gillian W. Shorter, and Mark D. Griffiths point in "An Empirical Study of Gender Differences in Online Gambling" (*Journal of Gambling Studies*, vol. 30, no. 1, March 2014) to research suggesting that Internet gambling may attract more females than offline gambling because of perceptions that it represents a "safer, less intimidating, anonymous, more fun and more tempting" alternative to gambling at casinos and other venues.

Gainsbury et al.'s survey of Australian adults finds that online gamblers are significantly more likely to play a wider variety of games than those who gamble exclusively offline and that they are significantly more likely to play table games. Internet gamblers also expressed a more benevolent view of gambling in general: only 33.3% of online gamblers believed that "the harm of gambling far outweighs the benefits," a belief held by 47% of the offline gamblers surveyed; and 8% of Internet gamblers believed that "the benefits somewhat outweigh the harm," a belief held by only 3.1% of offline gamblers.

THE EFFECTS OF ONLINE GAMBLING

Economics

As noted earlier, the economic effects of Internet gambling in the United States are highly dependent on the degree to which legalization spreads from state to state. Legal online gambling in Delaware, Nevada, and New Jersey generates state tax revenues that can be used for government programs. Operators of illegal online gambling sites, which are still believed to dominate the U.S. online gambling market and which are located offshore, pay no taxes to the federal, state, and local governments. Their revenues have little or no economic effect in the United States beyond the winnings of individual players.

The Department of Justice's 2011 opinion on the Wire Act leaves open the possibility for states that already allow other forms of gambling to proceed with the legalization of Internet gambling. This includes states where either tribal or commercial casino gambling is allowed, as well as states with lotteries or pari-mutuel racetrack gambling, so there is the potential for legal, taxable Internet gambling to spread to most states. In the 2015 edition of *Global Online Gambling*, MarketLine projects that the worldwide online gambling industry will grow by more than $20 billion in annual revenues between 2013 and 2018. This figure does not account for the possibility that legislation legalizing online gambling was under consideration in eight states as of August 2017. Should the U.S. online gambling industry fulfill such projections, it would generate substantial state and local tax revenues, as determined by the details of legislation in each state.

Although the early revenues generated in Delaware, Nevada, and New Jersey were generally modest, there is a possibility that the earning potential of Internet gambling could increase dramatically given the spread of legalization and multistate compacts allowing for games in which players from different states compete against one another. According to Demirjian, Nevada's online gambling revenues (which are limited to online poker) are expected to reach $50 million annually in the event that all or most states proceed with legalization.

A high proportion of the global online gambling industry's revenues comes from sports betting, the one form of online gambling that remains illegal in the United States. Thus, the spread of legalization in the United States is unlikely to include sports betting, so the legal U.S. Internet gambling industry, even in the most optimistic expansion scenarios, is unlikely to keep pace with the global industry.

Crime

MONEY LAUNDERING. Since Internet gambling operators first came into existence during the mid-1990s, law enforcement agencies have expressed concerns about the possible use of online gambling sites for money laundering. In testimony before the Senate Committee on Banking, Housing, and Urban Affairs, John G. Malcolm (March 18, 2003, http://banking.senate.gov/03_03hrg/031803/malcolm.htm), the former U.S. deputy assistant attorney general, explained that once the money has been stashed with an online casino, criminals can use the games themselves to transfer money to their associates. Some criminals set up private tables at online casino sites and then intentionally lose their money to business associates at the table. In other instances, the casino is part of the crime organization, so all the criminal has to do is lose money to the casino.

According to law enforcement officials, the factors that make online gambling susceptible to money laundering include the speed and anonymity with which financial transactions take place and the offshore locations of the gambling companies. In "Online Crime and Internet Gambling" (*Journal of Gambling Issues*, vol. 24, July 2010), John L. McMullan and Aunshul Rege document several ways that money laundering can take place through online gambling. For example, criminal networks can swamp online poker rooms with inferior "bots" and then play against these bots, allowing dirty money to be divided and exchanged through the online game. Bots can also be programmed to wager and fold to flood online games with illegal money. In such instances, criminals can take the last seat at the table, beat the bots, and cash out the winnings, thereby "cleaning" the money through the online poker room accounts.

CYBERNOMADS AND "DOT-CONS." The proliferation of online poker sites has led to increasingly sophisticated methods of "cheating" at the game. McMullan and Rege call these cheating individuals "cybernomads"—people who illegally make money from online poker sites, but not as part of criminal networks. The researchers explain that these players use artificial intelligence software that gives them a huge edge in online poker games. This software is used to monitor opponents to determine their betting patterns and styles and to calculate the odds. Other "cheaters" are hackers themselves who harvest personal information from online gambling sites and sell it on the black market.

McMullan and Rege note that criminals sometimes work in "dot-con" teams to cheat online poker sites. A notable case happened in 2007 during an online poker tournament at FullTiltPoker.com. During the middle of the tournament Chris "BluffMagCV" Vaughn sold his seat to Sorel "Imper1um" Mizzi, a more experienced player. As a result, the other tournament players were unable to recognize Mizzi's playing style and defeat him. However, Mizzi/Vaughn were denied the $1 million prize when FullTiltPoker.com discovered the deception. Dot-con teams also work together to modify game events or commands or to steal usernames, passwords, and account information.

Problem Gambling

Experts suggest that online gambling is more addictive than other types of gambling because of the ease of access, the unlimited frequency of play it allows, and the characteristics of certain online games, among other factors. Online gambling is quite different from traditional casino gambling because it is a solitary and anonymous activity that people can engage in, at any time of day or night, without being monitored. By contrast, casino gambling is a social activity and is subject to age restrictions and the possibility of self-exclusion. The fact that people can gamble online without allocating the same amount of time and effort to transportation and logistics further means that online gambling is likely used as a supplement to offline gambling, especially by those who are most at risk of problem and disordered gambling.

George T. Ladd and Nancy M. Petry of the University of Connecticut conducted a study of Internet gamblers and published their results in "Disordered Gambling among University-Based Medical and Dental Patients: A Focus on Internet Gambling" (*Psychology of Addictive Behaviors*, vol. 16, no. 1, March 2002). Between August 1999 and September 2000 the researchers surveyed 389 patients seeking free or reduced-cost services at the university's health and dental clinics. Using the South Oaks Gambling Screen (SOGS), a standard series of questions that are used to determine the probability that a person has a gambling problem (see Chapter 2), Ladd and Petry classified gamblers into three levels depending on their SOGS scores: Level 1 gamblers had SOGS scores of 0 to 2 and were considered not to have a gambling problem, Level 2 gamblers had SOGS scores of 3 to 4 and were considered probable problem gamblers, and Level 3 gamblers had SOGS scores of 5 or greater and were considered probable pathological gamblers. The mean (average) SOGS score of online gamblers was 7.8, compared with 1.8 for those who had no online gambling experience. Slightly more than 74% of Internet gamblers were rated at Levels 2 or 3, compared with 21.6% of the traditional gamblers.

Ladd and Petry then retrieved the medical records for those who participated in the study. The researchers find that Internet gamblers had poorer mental and physical health than did non-Internet gamblers. In detailing their study, Ladd and Petry state that "the availability of Internet gambling may draw individuals who seek out isolated and anonymous contexts for their gambling behaviors." Although problem gamblers are able to resist traveling to another state to play in casinos, Ladd and Petry determine that online gambling is more difficult to avoid because Internet sites are always open and accessible.

Petry reports the results of a follow-up study in "Internet Gambling: An Emerging Concern in Family Practice Medicine" (*Family Practice*, vol. 23, no. 4, August 2006). In this study, 1,414 adults in waiting areas of health clinics were given the SOGS test. Some 6.9% of adults reported ever gambling on the Internet, and 2.8% said they gambled online frequently. Of those who gambled frequently, nearly two-thirds (65.9%) were categorized as problem gamblers, as compared with 29.8% of those who reported ever gambling on the Internet and 7.6% of those who were classified as non-Internet gamblers.

Thomas Holtgraves of Ball State University finds in "Gambling, Gambling Activities, and Problem Gambling" (*Psychology of Addictive Behaviors*, vol. 23, no. 2, June 2009) that Internet gambling and betting on sports and horse races have higher "conversion rates" than other gambling activities; in other words, people who gamble on the Internet or bet on sports are more likely than other gamblers to continue doing so frequently. Holtgraves identifies the Internet's availability as a causal factor in this high conversion rate, stating, "Nothing is more available than the Internet; a player doesn't even need to leave home."

In "Gambling Involvement and Increased Risk of Gambling Problems" (*Journal of Gambling Studies*, vol. 29, no. 4, December 2013), James G. Phillips et al. revisit data from earlier studies and conduct new surveys in Australia in an attempt to determine whether the increased access to more forms of gambling offered by the Internet increases the risk of problem gambling and whether the use of newer and/or illegal gambling forms correlates with problem gambling. The researchers find that there was a significant correlation between the number of forms of gambling a person engaged in and the likelihood that he or she was a problem gambler. They further find that, among specific games, Internet poker and Internet sports wagering were most significantly associated with problem gambling. The fact that problem gamblers are involved in more forms of gambling and tend to shuttle between online and offline providers is noted by Phillips et al. as a significant new issue for attempts to treat and control problem gambling.

Underage Gambling

In January 2001 the American Psychiatric Association was already warning the public about Internet gambling. In a public health advisory, it noted that because online sites were not regulated, measures were not being taken to prevent underage gamblers from participating. It considered children, who already play nongambling games on the Internet, to be at significant risk of being lured to gambling sites. It also noted that few safeguards were in place to ensure the fairness of the Internet games or to establish exactly who had responsibility for operating them.

In "FTC Warns Consumers about Online Gambling and Children" (June 26, 2002, http://www.ftc.gov/news-events/press-releases/2002/06/ftc-warns-consumers-about-online-gambling-and-children), the Federal Trade Commission (FTC) also warns parents about children and online gambling. The FTC states "that minors can … access these sites easily, and that minors are often exposed to ads for online gambling on non-gambling websites." Furthermore, the commission notes that it examined more than 100 Internet gambling websites and found that 20% did not have any warnings directed toward children and that many lacked measures to block minors from gambling.

The Annenberg Public Policy Center at the University of Pennsylvania, which conducts the periodic National Annenberg Survey of Youth, has voiced similar concerns. The center states in "Internet Gambling Grows among Male Youth Ages 18 to 22" (October 14, 2010, http://www.annenbergpublicpolicycenter.org/Downloads/Releases/ACI/Card%20Playing%202010%20Release%20final.pdf) that 6.2% of male high school students between the ages of 14 and 17 gambled online at least once per month in 2010; in 2008 this figure was only 2.7%. At the same time, 1.5% of female high school students aged 14 to 17 years reported gambling online at least once per month in 2010, up from 0.5% in 2008. The numbers were significantly higher among older students, with 16% of college-aged males and 4.4% of college-aged females gambling online at least once per month in 2010.

In October 2011 Daniel Romer (http://www.annenbergpublicpolicycenter.org/Downloads/Adolescent_Risk/NASY/Dan%20Romer%20Testimony.pdf), the director of the Adolescent Communication Institute at the Annenberg Public Policy Center, testified before the U.S. House of Representatives Subcommittee on Commerce, Manufacturing, and Trade on the subject of online gambling. At the hearing, Romer argued that online gambling presented a particularly challenging situation because it involved illegal entities operating outside of the United States, making it more difficult for the U.S. government to implement legal safeguards that protect young people from engaging in this potentially self-destructive behavior. Nevertheless, he encouraged the adoption of "a new regime of online licensing and control" that would help "minimize the harm that this activity can inflict on the young and their families."

In "Mom, Dad It's Only a Game! Perceived Gambling and Gaming Behaviors among Adolescents and Young Adults: An Exploratory Study" (*International Journal of Mental Health and Addiction*, vol. 12, no. 6, December 2014), Filipa Calado, Joana Alexandre, and Mark D. Griffiths assess the relationship between gaming (i.e., video games) and gambling among Portuguese young people between the ages of 13 and 26 years. This is an issue of primary concern given the prevalence of video game play among young people and the numerous similarities between online gaming and online gambling, including the addictive qualities of both and the fact that video games can be played online for money. The researchers find that across the age spectrum of the sample, the young people considered gaming and gambling to be very similar activities and that when asked about the appropriate minimum age for gambling, most maintained that age is a less important consideration than one's level of skill at playing the game in question. According to Calado, Alexandre, and Griffiths, these findings suggest that both adolescents and young adults are subject to confusion about the difference between video games (which require and reward skill) and gambling (which are games of chance). These findings also suggest that a high level of familiarity with technology may be a risk factor for underage gambling and that young people who transition from video games to gambling may develop a false belief in their own skill, which is one of the hallmarks of problem gambling. Although adolescents in the sample group were in general more naive in their understanding of gambling than were the young adults, Calado, Alexandre, and Griffiths emphasize that "not a single participant expressed disapproval toward gambling."

IMPORTANT NAMES AND ADDRESSES

American Gaming Association
799 Ninth St. NW, Ste. 700
Washington, DC 20001
(202) 552-2675
FAX: (202) 552-2676
URL: https://www.americangaming.org/

Annenberg Public Policy Center
202 S. 36th St.
Philadelphia, PA 19104-3806
(215) 898-9400
FAX: (215) 573-7116
URL: http://www.annenbergpublic policycenter.org/

Arizona Department of Gaming / Racing Division
1110 W. Washington, Ste. 450
Phoenix, AZ 85007
(602) 771-4263
FAX: (602) 255-3883
URL: https://racing.az.gov/

Arizona Lottery
4740 E. University Dr.
Phoenix, AZ 85034
(480) 921-4400
URL: https://www.arizonalottery.com/

Arkansas Department of Finance and Administration Arkansas Racing Division
1515 Bldg.
1515 W. Seventh St., Ste. 505
Little Rock, AR 72201
(501) 682-1467
FAX: (501) 682-5273
URL: http://www.dfa.arkansas.gov/offices/racingCommission/Pages/default.aspx

Arkansas Scholarship Lottery
PO Box 3238
Little Rock, AR 72203-3239
(501) 683-2000
E-mail: aslinfo@arkansas.gov
URL: http://myarkansaslottery.com/

California Horse Racing Board
1010 Hurley Way, Ste. 300
Sacramento, CA 95825
(916) 263-6000
1-800-805-7223
FAX: (916) 263-6042
E-mail: TIPS@chrb.ca.gov
URL: http://www.chrb.ca.gov/

California Lottery
4106 East Commerce Way, Ste. 100
Sacramento, CA 95834
(916) 830-0292
URL: http://www.calottery.com/

Center for Gaming Research
University of Nevada, Las Vegas
4505 Maryland Pkwy.
Box 457010
Las Vegas, NV 89154-7010
(702) 895-2242
FAX: (702) 895-2253
E-mail: dgs@unlv.nevada.edu
URL: http://gaming.unlv.edu/

Colorado Division of Gaming
Colorado Department of Revenue
17301 W. Colfax Ave., Ste. 135
Golden, CO 80401
(303) 205-1300
FAX: (303) 205-1342
E-mail: dor_gamingweb@state.co.us
URL: https://www.colorado.gov/enforcement/gaming

Colorado Lottery
225 N. Main St.
Pueblo, CO 81003
(719) 546-2400
FAX: (719) 546-5208
URL: https://www.coloradolottery.com/

Connecticut Department of Consumer Protection, Gaming Division
450 Columbus Blvd.
Hartford, CT 06103
(860) 713-6100
URL: http://www.ct.gov/dcp/cwp/view.asp?a=4107&q=480854

Connecticut Lottery Corporation
777 Brook St.
Rocky Hill, CT 06067
(860) 713-2000
FAX: (860) 713-2805
E-mail: ctlottery@ctlottery.org
URL: http://www.ctlottery.org/

DC Office of Lottery and Charitable Games
2235 Shannon Place SE, Rm. 5007
Washington, DC 20020
(202) 645-8000
URL: http://lottery.dc.gov/

Delaware Lottery
McKee Business Park
1575 McKee Rd., Ste. 102
Dover, DE 19904
(302) 739-5291
FAX: (302) 739-6706
URL: http://www.delottery.com/

Florida Department of Business and Professional Regulation Division of Pari-mutuel Wagering
2601 Blair Stone Rd.
Tallahassee, FL 32399-1035
(850) 487-1395
FAX: (850) 488-0550
URL: http://www.myfloridalicense.com/dbpr/pmw/index.html

Florida Lottery
250 Marriott Dr.
Tallahassee, FL 32301
(850) 487-7787
FAX: (850) 488-8049
URL: http://www.flalottery.com/

Gamblers Anonymous
PO Box 17173
Los Angeles, CA 90017

(626) 960-3500
FAX: (626) 960-3501
URL: http://www.gamblersanonymous.org/

Georgia Lottery Corporation
250 Williams St., Ste. 3000
Atlanta, GA 30303
(404) 215-5000
E-mail: glottery@galottery.org
URL: https://www.galottery.com/

Hoosier Lottery
1302 N. Meridian St.
Indianapolis, IN 46202
1-800-955-6886
URL: https://www.hoosierlottery.com/

Idaho Lottery
1199 Shoreline Ln., Ste. 100
Boise, ID 83702
(208) 334-2600
E-mail: info@idaholottery.com
URL: http://www.idaholottery.com/

Idaho State Racing Commission
700 S. Stratford Dr.
Meridian, ID 83642
(208) 884-7080
FAX: (208) 884-7098
URL: https://isp.idaho.gov/racing/

Illinois Gaming Board
801 S. Seventh St., Ste. 400—South
Springfield, IL 62703
(217) 524-0226
URL: http://www.igb.illinois.gov/

Illinois Lottery
101 W. Jefferson St.
Springfield, IL 62702
(217) 524-6435
E-mail: LotteryInfo@NorthstarLottery.net
URL: http://www.illinoislottery.com/

Illinois Racing Board
100 W. Randolph St., Ste. 5-700
Chicago, IL 60601
(312) 814-2600
URL: https://www.illinois.gov/irb/Pages/default.aspx

Indiana Gaming Commission
East Tower, Ste. 1600
101 W. Washington St.
Indianapolis, IN 46204
(317) 233-0046
FAX: (317) 233-0047
URL: http://www.in.gov/igc/

Indiana Horse Racing Commission
Hoosier Park
4500 Dan Patch Circle
Anderson, IN 46013
(765) 609-4855
URL: http://www.in.gov/hrc/

Institute for the Study of Gambling and Commercial Gaming
University of Nevada–Reno
1664 N. Virginia St.
Reno, NV 89557-0025
(775) 784-6850
FAX: (775) 784-1057
URL: https://www.unr.edu/gaming/

Iowa Lottery
13001 University Ave.
Clive, IA 50325-8225
(515) 725-7900
URL: http://www.ialottery.com/

Iowa Racing and Gaming Commission
DMACC Capitol Center
1300 Des Moines St., Ste. 100
Des Moines, IA 50309
(515) 281-7352
E-mail: irgc@iowa.gov
URL: https://irgc.iowa.gov/

Jockey Club
40 E. 52nd St.
New York, NY 10022
(212) 371-5970
FAX: (212) 371-6123
URL: http://www.jockeyclub.com/

Kansas Lottery
128 N. Kansas Ave.
Topeka, KS 66603
(785) 296-5700
E-mail: lotteryinfo@kslottery.net
URL: http://www.kslottery.com/

Kansas Racing and Gaming Commission
700 SW Harrison, Ste. 500
Topeka, KS 66603-3754
(785) 296-5800
FAX: (785) 296-0900
E-mail: krgc@krgc.ks.gov
URL: http://krgc.ks.gov/

Kentucky Horse Racing Commission
4063 Iron Works Pkwy., Bldg. B
Lexington, KY 40511
(859) 246-2040
FAX: (859) 246-2039
URL: http://khrc.ky.gov/Pages/default.aspx

Kentucky Lottery
1011 W. Main St.
Louisville, KY 40202
(502) 583-2000
1-877-789-4532
E-mail: help@kylottery.com
URL: https://www.kylottery.com/

Louisiana Gaming Control Board
7901 Independence Blvd., Bldg. A
Baton Rouge, LA 70806
(225) 925-1846
1-888-295-8450
FAX: (225) 925-1945
E-mail: LGCB@la.gov
URL: http://lgcb.dps.louisiana.gov/

Louisiana Lottery Corporation
555 Laurel St.
Baton Rouge, LA 70801
(225) 297-2000
info@louisianalottery.com
URL: https://louisianalottery.com/

Maine Gambling Control Unit
Department of Public Safety
45 Commerce Dr.
SHS#87
Augusta, ME 04333-0087
(207) 626-3900
FAX: (207) 287-4356
E-mail: Gambling.Board@maine.gov
URL: http://www.maine.gov/dps/GambBoard/index.html

Maine Lottery
Bureau of Alcoholic Beverages and Lottery Operations
Eight State House Station
Augusta, ME 04333-0008
(207) 287-3721
1-800-452-8777
FAX: (207) 287-6769
MaineLottery@Maine.gov
URL: http://www.mainelottery.com/

Maine State Harness Racing Commission
Maine Department of Agriculture, Conservation, and Forestry
28 State House Station
Augusta, ME 04333-0022
(207) 287-3221
FAX: (207) 287-7548
URL: http://www.maine.gov/dacf/harnessracing/index.shtml

Maryland Center of Excellence on Problem Gambling
University of Maryland School of Medicine
Waterloo Crossing
5900 Waterloo Rd., Ste. 200
Columbia, MD 20145
(667) 214-2120
E-mail: dgaspar@psych.umaryland.edu
URL: http://www.mdproblemgambling.com/

Maryland Racing Commission
300 E. Towsontowne Blvd.
Towson, MD 21286
(410) 296-9682
FAX: (410) 296-9687
E-mail: dloplmarylandracingcommission-dllr@maryland.gov
URL: http://www.dllr.state.md.us/racing/

Maryland State Lottery Agency
Montgomery Business Park
1800 Washington Blvd., Ste. 330
Baltimore, MD 21230
(410) 230-8800
URL: http://www.mdlottery.com/

Massachusetts Gaming Commission
101 Federal St., 12th Floor
Boston, MA 02110

(617) 979-8400
FAX: (617) 725-0258
E-mail: mgccomments@state.ma.us
URL: http://www.massgaming.com/

Massachusetts State Lottery Commission
60 Columbian St.
Braintree, MA 02184
(781) 849-5555
URL: http://www.masslottery.com/

Michigan Gaming Control Board
3062 W. Grand Blvd., Ste. L-700
Detroit, MI 48202-6062
(313) 456-4100
FAX: (313) 456-4200
E-mail: MGCBweb@michigan.gov
URL: http://www.michigan.gov/mgcb

Michigan Lottery
101 E. Hillsdale St.
Lansing, MI 48909
(517) 335-5600
URL: http://www.michigan.gov/lottery

Minnesota Gambling Control Board
1711 W. County Rd. B, Ste. 300 S.
Roseville, MN 55113
(651) 539-1900
FAX: (651) 639-4032
URL: http://www.mn.gov/gcb/

Minnesota Lottery
2645 Long Lake Rd.
Roseville, MN 55113
(651) 635-8273
E-mail: lottery@mnlottery.com
URL: http://www.mnlottery.com/

Minnesota Racing Commission
15201 Zurich St. NE, Ste. 212
Columbus, MN 55025
(651) 925-3951
FAX: (651) 925-3953
URL: https://www.mrc.state.mn.us/index.htm

Mississippi Gaming Commission
620 North St., Ste. 200
Jackson, MS 39202
(601) 576-3800
1-800-504-7529
FAX: (601) 576-3929
E-mail: info@mgc.state.ms.us
URL: http://www.msgamingcommission.com/

Missouri Gaming Commission
3417 Knipp Dr.
PO Box 1847
Jefferson City, MO 65102
(573) 526-4080
FAX: (573) 526-1999
E-mail: PublicRelation@mgc.dps.mo.gov
URL: http://www.mgc.dps.mo.gov/

Missouri Lottery
1823 Southridge Dr.
PO Box 1603
Jefferson City, MO 65109-1603
(573) 751-4050
FAX: (573) 751-5188
URL: http://www.molottery.com/

Montana Gambling Control Division
Department of Justice
2550 Prospect Ave.
PO Box 201424
Helena, MT 59620-1424
(406) 444-1971
URL: https://dojmt.gov/gaming/

Montana Lottery
2525 N. Montana Ave.
Helena, MT 59601-0598
(406) 444-5825
E-mail: montanalottery@mt.gov
URL: https://www.montanalottery.com/

National Association of Fundraising Ticket Manufacturers
180 E. Fifth St., Ste. 940
St. Paul, MN 55101
(651) 644-4710
FAX: (651) 644-5904
URL: http://www.naftm.org/

National Center for Responsible Gaming
900 Cummings Center, Ste. 321-U
Beverly, MA 01915
(978) 338-6610
FAX: (978) 552-8452
E-mail: info@ncrg.org
URL: http://www.ncrg.org/

National Council on Problem Gambling
730 11th St. NW, Ste. 601
Washington, DC 20001
(202) 547-9204
FAX: (202) 547-9206
E-mail: ncpg@ncpgambling.org
URL: http://www.ncpgambling.org/

National Indian Gaming Association
224 Second St. SE
Washington, DC 20003
(202) 546-7711
E-mail: questions@indiangaming.org
URL: http://www.indiangaming.org/

National Indian Gaming Commission
90 K St. NE, Ste. 200
Washington, DC 20002
(202) 632-7003
FAX: (202) 632-7066
E-mail: contactus@nigc.gov
URL: https://www.nigc.gov/

Nebraska Lottery
137 NW 17th St.
PO Box 98901
Lincoln, NE 68528-1204
(402) 471-6100
E-mail: lottery@nelottery.com
URL: https://nelottery.com/

Nebraska State Racing Commission
5903 Walker Ave.
Lincoln, NE 68507
(402) 471-4155
E-mail: diane.vandeun@nebraska.gov
URL: http://nebraskaracingcommission.com/

Nevada Gaming Commission and Gaming Control Board
1919 College Pkwy.
Carson City, NV 89706
(775) 684-7750
FAX: (775) 687-5817
URL: http://gaming.nv.gov/

New Hampshire Lottery Commission
14 Integra Dr.
Concord, NH 03301
(603) 271-3391
1-800-852-3324
FAX: (603) 271-1160
E-mail: webmaster@Lottery.nh.gov
URL: http://www.nhlottery.com/

New Hampshire Lottery Commission Racing and Charitable Gaming Commission
14 Integra Dr.
Concord, NH 03301
(603) 271-3391
FAX: (603) 271-1160
E-mail: RCGDivision@lottery.nh.gov
URL: https://www.racing.nh.gov/

New Jersey Casino Control Commission
Arcade Bldg.
Tennessee Ave. and Boardwalk
Atlantic City, NJ 08401
(609) 441-3422
E-mail: communications@ccc.state.nj.us
URL: http://www.state.nj.us/casinos/

New Jersey Lottery
Lawrence Park Complex
1333 Brunswick Avenue Circle
Trenton, NJ 08648
(609) 599-5800
FAX: (609) 599-5935
E-mail: publicinfo@lottery.nj.gov
URL: https://www.njlottery.com/

New Jersey Racing Commission
140 E. Front St., Third Floor
Trenton, NJ 08625
(609) 292-0613
FAX: (609) 599-1785
URL: http://www.nj.gov/oag/racing/index.html

New Mexico Gaming Control Board
4900 Alameda Blvd. NE
Albuquerque, NM 87113
(505) 841-9700
FAX: (505) 841-9725
URL: http://www.nmgcb.org/

New Mexico Lottery
4511 Osuna Rd. NE
Albuquerque, NM 87109
(505) 342-7600
URL: http://www.nmlottery.com/

New Mexico Racing Commission
4900 Alameda NE
Albuquerque, NM 87113
(505) 222-0700
FAX: (505) 222-0713
E-mail: rc.info@state.nm.us
URL: http://nmrc.state.nm.us

New York State Gaming Commission
PO Box 7500
Schenectady, NY 12301-7500
(518) 388-3300
E-mail: info@gaming.ny.gov
URL: https://gaming.ny.gov/

North American Association of State and Provincial Lotteries
7470 Auburn Rd., LL1
Concord, OH 44077
(440) 361-7962
E-mail: info@nasplhq.org
URL: http://www.naspl.org/

North Carolina Education Lottery
2728 Capital Blvd., Ste. 144
Raleigh, NC 27604
(919) 715-6886
FAX: (919) 715-8833
URL: http://www.nc-educationlottery.org/

North Dakota Lottery
1050 E. Interstate Ave., Ste. 200
Bismarck, ND 58503-5574
(701) 328-1574
1-877-635-6886
E-mail: ndlottery@nd.gov
URL: http://www.ndlottery.org/

North Dakota Office of Attorney General Gaming Division
600 E. Boulevard Ave., Dept. 125
Bismarck, ND 58505
(701) 328-4848
FAX: (701) 328-3535
URL: https://attorneygeneral.nd.gov/licensing-and-gaming/gaming

Ohio Casino Control Commission
10 W. Broad St., Sixth Floor
Columbus, OH 43215
(614) 387-5858
1-855-800-0058
FAX: (614) 485-1007
E-mail: info@casinocontrol.ohio.gov
URL: http://casinocontrol.ohio.gov/

Ohio Lottery Commission
615 W. Superior Ave.
Cleveland, OH 44113
1-800-686-4208
URL: http://www.ohiolottery.com/

Ohio State Racing Commission
77 S. High St., 18th Floor
Columbus, OH 43215
(614) 466-2757
FAX: (614) 466-1900
URL: http://www.racing.ohio.gov/

Oklahoma Horse Racing Commission
Shepherd Mall
2800 N. Lincoln Blvd.
Oklahoma City, OK 73105
(405) 943-6472
FAX: (405) 943-6474
URL: http://www.ohrc.org/

Oregon Lottery
500 Airport Rd. SE
Salem, OR 97301
(503) 540-1000
FAX: (503) 540-1001
E-mail: lottery.webcenter@state.or.us
URL: http://www.oregonlottery.org/

Oregon Racing Commission
800 NE Oregon St., Ste. 310
Portland, OR 97232
(971) 673-0207
URL: http://www.oregon.gov/RACING/pages/index.aspx

Pennsylvania Gaming Control Board
PO Box 69060
Harrisburg, PA 17106-9060
(717) 346-8300
FAX: (717) 346-8350
E-mail: pgcb@pa.gov
URL: http://gamingcontrolboard.pa.gov/

Pennsylvania Lottery
1200 Fulling Mill Rd., Ste. 1
Middletown, PA 17057
(717) 702-8000
1-800-692-7481
FAX: (717) 702-8024
URL: https://www.palottery.state.pa.us/

Pew Research Center
1615 L St. NW, Ste. 800
Washington, DC 20036
(202) 419-4300
FAX: (202) 419-4349
URL: http://www.pewresearch.org/

Rhode Island Lottery
1425 Pontiac Ave.
Cranston, RI 02920
(401) 463-6500
FAX: (401) 463-5669
URL: http://www.rilot.com/

South Carolina Education Lottery
PO Box 11949
Columbia, SC 29211-1949
(803) 737-4419
URL: http://www.sceducationlottery.com/

South Dakota Commission on Gaming
445 E. Capitol Ave.
Pierre, SD 57501-3185
(605) 773-6050
FAX: (605) 773-6053
URL: http://dor.sd.gov/gaming/

South Dakota Lottery
711 E. Wells Ave.
Pierre, SD 57501
(605) 773-5770
FAX: (605) 773-5786
E-mail: lottery@state.sd.us
URL: https://lottery.sd.gov/

Stop Predatory Gambling Foundation
100 Maryland Ave. NE, Rm. 310
Washington, DC 20002
(202) 567-6996
E-mail: mail@stoppredatorygambling.org
URL: http://stoppredatorygambling.org/

Tennessee Lottery
One Century Place
26 Century Blvd., Ste. 200
Nashville, TN 37214
(615) 324-6500
URL: http://www.tnlottery.com/

Texas Lottery Commission
611 E. Sixth St.
Austin, TX 78701
(512) 344-5000
1-800-375-6886
FAX: (512) 344-5080
E-mail: customer.service@lottery.state.tx.us
URL: http://www.txlottery.org/

Texas Racing Commission
8505 Cross Park Dr., Ste. 110
Austin, TX 78754
(512) 833-6699
FAX: (512) 833-6907
URL: http://www.txrc.texas.gov/

Vermont Lottery Commission
1311 U.S. Rte. 302
Barre, VT 05641
(802) 479-5686
1-800-322-8800
FAX: (802) 479-4294
E-mail: staff@vtlottery.com
URL: http://www.vtlottery.com/

Virginia Lottery
600 E. Main St.
Richmond, VA 23219
(804) 692-7000
FAX: (804) 692-7102
URL: http://www.valottery.com/

Virginia Racing Commission
5707 Huntsman Rd., Ste. 201-B
Richmond, VA 23250
(804) 966-7400
E-mail: Kimberly.Mackey@VRC.Virginia.gov
URL: http://www.vrc.virginia.gov/

Washington Horse Racing Commission
6326 Martin Way E., Ste. 209
Olympia, WA 98516
(360) 459-6462
FAX: (360) 459-6461
E-mail: whrc@whrc.state.wa.us
URL: http://www.whrc.wa.gov/

Washington's Lottery
814 Fourth Ave. E.
Olympia, WA 98506
(360) 664-4720
FAX: (360) 515-0416
E-mail: director@walottery.com
URL: http://www.walottery.com/

Washington State Gambling Commission
4565 Seventh Ave. SE
Lacey, WA 98503
(360) 486-3440
1-800-345-2529
FAX: (360) 486-3629
E-mail: askus@wsgc.wa.gov
URL: http://www.wsgc.wa.gov/

West Virginia Lottery
900 Pennsylvania Ave.
Charleston, WV 25302
(304) 558-0500
1-800-982-2274
FAX: (304) 558-3321
E-mail: mail@wvlottery.com
URL: http://www.wvlottery.com/

West Virginia Racing Commission
900 Pennsylvania Ave., Ste. 533
Charleston, WV 25302
(304) 558-2150
FAX: (304) 558-6319
URL: http://www.racing.wv.gov/Pages/default.aspx

Wisconsin Division of Gaming
3319 W. Beltline Hwy., Ste. 1
Madison, WI 53713
(608) 270-2533
URL: http://www.doa.state.wi.us/Divisions/Gaming

Wisconsin Lottery
PO Box 8941
Madison, WI 53708-8941
(608) 261-4916
E-mail: info@wilottery.com
URL: http://www.wilottery.com/

Wyoming Lottery Corporation
1620 Central Ave., Ste. 100
Cheyenne, WY 82001
(307) 432-9300
1-855-995-6886
URL: http://www.wyolotto.com/

Wyoming Pari-mutuel Commission
Energy II Bldg., Ste. 335
951 Werner Ct.
Casper, WY 82601
(307) 265-4015
FAX: (307) 265-4279
URL: http://parimutuel.state.wy.us

RESOURCES

Several resources useful to this book were published by companies and organizations within the gambling industry. The most notable are *2016 State of the States: The AGA Survey of the Casino Industry* (November 2016) by the American Gaming Association, *Gambling Impact Study* (October 2013) by the Spectrum Gaming Group, and *NAFTM 2015 Annual Report* (2016) by the National Association of Fundraising Ticket Manufacturers.

Each state that allows gambling releases data and/or issues monthly, quarterly, or annual reports that describe revenues, employment, tax payments, and more. Reports and data sets helpful for this book included *Gaming Revenue Report: December 31, 2016* (2017) by the Nevada State Gaming Control Board; *2015 Annual Report* (2016) by the New Jersey Casino Control Commission; *Quarterly Reports* (2017) of the Mississippi Gaming Commission; *21st Annual Report to the Louisiana State Legislature* (2017) by the Louisiana Gaming Control Board; *Annual Report to Governor Mike Pence* (September 2016) by the Indiana Gaming Commission; *2016 Annual Report* (2017) of the Illinois Gaming Board; *2016 Michigan Gaming Control Board Annual Report* (April 2017) by the Michigan Gaming Control Board; *Pennsylvania Control Board 2015–2016 Annual Report* (2017) by the Pennsylvania Gaming Control Board; and numerous other items of information made available on the websites of individual state gambling regulators.

The National Indian Gaming Commission releases data on tribal gambling revenues at the "Gaming Revenue Reports" page of its website. The National Indian Gaming Association's website also provides an assortment of valuable statistics and other research pertaining to tribal gaming.

The North American Association of State and Provincial Lotteries provides data relating to lottery revenues at its website, and the individual state lotteries each provide data at their individual websites. Particularly helpful in providing a picture of lottery player demographics was the Texas Lottery Commission's *Demographic Survey of Texas Lottery Players 2016* (December 2016).

Organizations devoted to problem gambling that provided helpful information include Gamblers Anonymous, the National Center for Responsible Gaming, and the National Council on Problem Gambling. Most states run their own programs to address problem gambling. One helpful report was *Expanding the Vision: FY 2016 Annual Report* (2016) by the Maryland Center of Excellence on Problem Gambling, which provided useful data on problem gamblers.

Additional key insights about problem gamblers and the psychology of gambling came from scholarly journals, particularly the *Journal of Gambling Studies*, many issues of which were consulted in the compiling of this book. Other scientific and educational publications consulted include *Journal of Gambling Issues*, published by the Centre for Addiction and Mental Health in Toronto, Canada; *American Journal of Geriatric Psychiatry*; *American Journal of Psychiatry*; *American Journal on Addictions*; *Applied Economics*; *Behaviour Research and Therapy*; *Econ Journal Watch*; *Journal of Affective Disorders*; *Psychiatric Times*; *Psychology of Addictive Behaviors*; and *Review of Economics and Statistics*.

Other key data for the book came from Gallup, Inc., the Annenberg Public Policy Center at the University of Pennsylvania, the Center for Gaming Research at the University of Nevada, Las Vegas, the Pew Research Center, the U.S. Census Bureau, the U.S. Bureau of Labor Statistics, and the Federal Bureau of Investigation.

INDEX

Page references in italics refer to photographs. References with the letter t *following them indicate the presence of a table. The letter* f *indicates a figure. If more than one table or figure appears on a particular page, the exact item number for the table or figure being referenced is provided.*

A

Abdollahnejad, Mohammad Reza, 66
About Half of Americans Play State Lotteries (Auter), 6–7
"About Mohegan Sun" (Mohegan Sun), 53
"About Us" (Foxwoods Casino and Resort), 53
Abramoff, Jack, 71–72
Addiction
- Gamblers Anonymous 12-step recovery program, 19*t*
- online gambling, 108–109
- questions for determining if a person is a compulsive gambler, 18*t*

Adelson, Sheldon, 58, 102
Adolescent Communication Institute, 109
Adolescents
- gambling among, 16–17
- identified as pathological gamblers, 18
- Internet games, underage gambling and, 109–110
- *See also* Young people

AGA. *See* American Gaming Association
"AGA Code of Conduct for Responsible Gaming" (AGA), 68
"Agreed to Civil Fine Assessment" (NGIC), 47
AGTOA (American Greyhound Track Operators Association), 93
Agua Caliente Band of Cahuilla Indians, 55
Alexandre, Joana, 109–110
Allen, Jeff, 20
Allis, Kevin J., 63
"Amantadine in the Treatment of Pathological Gambling: A Case Report" (Pettorruso et al.), 20
American Gaming Association (AGA)
- on acceptability of gambling, 57
- "AGA Code of Conduct for Responsible Gaming," 68
- on California tribal casinos, 55
- on casino slot machines, 22
- on casino visitors, activities participated in by, 23–24
- on casino visitors, numbers of, 63
- on casinos, acceptability of, 22
- on commercial casino revenues, 21, 45, 58
- on commercial casino tax revenues, 59
- on commercial casinos, numbers operating in U.S., 25
- on commercial casinos, revenues generated by, 9
- on commercial casinos operating in Nevada, 25–26
- on gamblers budget limits, 67
- on gambling activity, prevalence rates, 16
- on house advantage, 1, 23
- on illegal gambling operators, prosecution of, 5
- on illegal sports gambling, monies wagered on, 97
- on Illinois casino market, 36
- on Internet wagering, numbers participating in, 106–107
- on Las Vegas strip/downtown casino revenues, 26–27
- on lottery tickets, number of purchasers, 78
- on Louisiana casino market, 36
- on Mississippi Gulf Coast casino market, 34
- on numbers employed in commercial casinos, 59–60
- online gambling, position on, 103
- on states allowing commercial casinos, 4

American Greyhound Council, 93
American Greyhound Track Operators Association (AGTOA), 93
American Indian Rights and Resource Organization, 55
American Pharoah, 90
American Psychiatric Association (APA)
- on compulsive gambling, 6
- on criteria for gambling disorder, lottery players and, 85
- online gambling, warning about, 109
- on pathological gambling, 17

American Quarter Horse Association, 90
American Society for the Prevention of Cruelty to Animals (ASPCA), 93
Americans Hold Record Liberal Views on Most Moral Issues (Jones), 6–7
Americans with Disabilities Act, 63
An Analysis of Internet Gambling and Its Policy Implications (Stewart), 101–102
Animal fighting
- cockfighting, 99
- dogfighting, 99
- gambling on, 98–99

Annapolis Jockey Club, 89
Anne, Queen of England, 2
Annenberg Adolescent Communication Institute, 16
Annenberg Public Policy Center
- on Internet gambling among youth, 109
- on young people, gambling among, 16

Annual Report 2011 (New Jersey Casino Control Commission), 65
Annual Report 2016 (Missouri Gaming Commission), 68
Annual Report to Governor Mike Pence (Indiana Gaming Commission), 33
Antidepressants, 20
Antze, Paul, 19
APA. *See* American Psychiatric Association
Apache tribal grouping, 45

"Approved Management Contracts" (NIGC), 51
Arabian horses, 90
Arabian Jockey Club, 90
Ariyabuddhiphongs, Vanchai
on lotteries, at-risk gamblers and, 85
on lotteries, uniqueness of, 76
on lottery players, demographics of, 78
on playing lottery, reasons for, 80–81
Arrests
for animal fighting, 99
for illegal gambling, rarity of, 12
of minors in casinos, 68
Missouri casino, by month, 71(*t*6.5)
by type of crime, estimated number, 13*t*
ASPCA (American Society for the Prevention of Cruelty to Animals), 93
Associated Press (AP)
"Experts: Online Gambling Slowed by Illegal Action," 106
"Landless Tribe Waits Federal Recognition," 48
Association of Gaming Equipment Manufacturers, 10
Atlantic City, New Jersey
casino hotels, opening of in, 4
casino industry, by source of revenue, 30(*t*4.5)
casinos of, 28–29
casinos of, effects on casino operations nationally, 21
corporate casino, first opened in, 9
economic effects of casino industry on, 61
tourism in, 64–65
Atlantic City Gaming Revenue: Annual Statistics for Total, Slot, Table, & Internet Win, 1978–2016 (Center for Gaming Research), 104–105
At-risk gamblers
definition of, 17
lottery availability and, 85
Australia, online gambling in, 107
Auter, Zac, 6–7
"Average/Median Price Per Yearling" (Jockey Club), 92

B

Bad Bet on the Bayou: The Rise of Gambling in Louisiana and the Fall of Governor Edwin Edwards (Bridges), 35
Banked games, 22
Bankruptcy, gambling and, 66
Bay Mills Indian Community
King's Club opening by, 46
off-reservation casino, legality of, 52
Bay Mills Indian Community, Michigan v., 52
Benedict, Jeff, 53
Benston, Liz, 24
Bet Smart campaign, 6
Betting
activities, skill *vs.* chance, 1
fantasy sports, 96–97
Greyhound racing, 93
illegal sports gambling, monies wagered on, 97
parlay betting, 93
sports, overview of, 95
See also Gambling
Betting limits, 43
Betting pool
horse racing, overview of, 90–91
pari-mutuel gambling, 89
BIA (Bureau of Indian Affairs)
federal recognition of tribes, 45
functions of, 47–48
Bible, 1
The Big Spin (television show), 76
Bingo
charity gambling and, 13
halls, history of tribal, 45–46
halls, Native American operated, 5
San Manuel Indian Bingo and Casino, 63
Black, Donald W.
"Extended Release Carbamazepine in the Treatment of Pathological Gambling: An Open-Label Study," 20
"Suicide Ideations, Suicide Attempts, and Completed Suicide in Persons with Pathological Gambling and Their First-Degree Relatives," 66
Blackfeet Tribe of the Blackfeet Indian Reservation, 45
Blevins, Audie, 43
Blume, Sheila B., 17
Bookmaking
illegal sports, effects on society, 99
illegal sports betting and, 98
organized crime involvement in, 12
in sports betting, overview of, 94
Bots, 108
Boulder Dam, 3
Boyd Gaming Corporation, 27
Breakage, 89
Breeding, horse, 92
Brennan, John, 94
Bridges, Tyler, 35
Brooks, Robert B., 20
Brown, Meredith, 20
Brustein, Joshua, 96
Bullbaiting, 2
Bureau of Indian Affairs (BIA)
federal recognition of tribes, 45
functions of, 47–48
"Bureau of Indian Affairs Proposes Revising Rules for Recognizing Native American Tribes" (Melia), 48
Burk, Martha Jane, 43
Bush, George W., 52
Businesses, small, 11
Butterworth, Seminole Tribe of Florida v., 45

C

Cabazon Band of Mission Indians, California v., 5, 46
Calado, Filipa, 109–110
Calamity Jane, 43
Calcagno, Peter T., 69
Calcutta pools, 95
California
economic output of tribal casinos, 59
government finances, 85*t*
horse racetracks, closing of, 92
horse racetracks in, revenues of, 89
racinos, disallowing of in, 91
tribal casinos, 55–56
tribal gaming in, 5
California Federation of Teachers, 55
California Horse Racing Board, 89
California Nations Indian Gaming Association
on California tribal casinos, revenues of, 59
on tribal gaming, economic impact on state, 56
California State Lottery
The Big Spin, 76
educational funding of, 84
group play, 80
"Historical Lucky Retailers," 77
on unclaimed winnings, 84
California Tax Report Association, 55
"California Tribal Communities" (Judicial Branch of California), 55
California v. Cabazon Band of Mission Indians, 5, 46
Capone, Al, 98
Card games
casino *vs.* online, 103
gambling classes and, 46
in medieval times, 2
Careers, casino, 59–63
See also Employment
Caribbean, Internet gambling sites in, 5
The Casino Career Institute, 63
Casino City Press, 10
Casino owners/operators
Boyd Gaming Corporation, 10
Caesars Entertainment Corporation, 10
MGM Resorts International, 10
overview of, 9–10
Penn National Gaming, 11
Red Rock Resorts, 11
Casino Reinvestment Development Authority (CRDA), 64–65
Casinonomics: The Socioeconomic Impacts of the Casino Industry (Walker), 59, 61
Casinos
acceptability of, 22
casino gaming revenues, by state, 10*t*
debate over, pros/cons, 59, 61
games offered by, 22–23

- growth of in U.S., 5
- historical/current status of, 21–22
- house advantage of, by game, 23*t*
- legalized in Nevada, 3
- methods for encouraging gambling in, 23–24
- overview of, 21
- statistical advantage of, 1
- tax revenues generated by casino gambling, by state, 15*t*

Casinos, commercial
- Atlantic City casino industry employment figures, by casino, 32(*t*4.7)
- Atlantic City casino industry revenues, by source of revenue, 30(*t*4.5)
- Colorado, 41
- Colorado casino revenues, 42(*t*4.23)
- consumer spending at, 14(*t*2.4)
- Detroit, Michigan, casino revenues, 40(*t*4.20)
- Detroit, Michigan, casino state taxes, 40(*t*4.21)
- Illinois, 36–37
- Illinois gaming revenues, 38(*t*4.16)
- Illinois gaming revenues, by casino/type, 38(*t*4.17)
- Illinois gaming taxes, state *vs.* local share of, 39(*t*4.18)
- Indiana, 33
- Indiana casino revenues/taxes, 33*t*
- Iowa, 40–41
- Iowa casino financial data, 41*t*
- Kansas, 43
- Las Vegas downtown gaming revenue, by type, 29*t*
- Las Vegas strip gaming revenue, by type, 28*t*
- Louisiana, 35–36
- Louisiana land-based casino financial data, 37(*t*4.14)
- Louisiana riverboat gaming statistics, 36*t*
- Louisiana video gaming statistics, 38(*t*4.15)
- Maine, 43
- Maryland, 42–43
- Michigan, 39–40
- Mississippi, 33–35
- Mississippi casino revenues, by region, 34*t*
- Mississippi casino tables hold, by game/region, 35*t*
- Missouri, 37–39
- Missouri casino gaming financial data, by casino location, 39(*t*4.19)
- Nevada, 25–28
- Nevada gaming revenue, 26*f*
- Nevada gaming revenue, by type, 27*t*
- New Jersey, 28–29
- New Jersey casino gaming revenue, 30(*t*4.4)
- New Jersey casino gaming revenue, by type, 31*t*
- numbers employed in, 59–60
- Ohio, 41–42
- online gambling, backing of, 103
- overview of, 25
- Pennsylvania, 31–32
- Pennsylvania casino revenues, 32(*t*4.8)
- political contributions of, 70
- riverboat casino financial data, by state, 37(*t*4.13)
- South Dakota, 43
- South Dakota casino revenues, 44*f*
- tribal-commercial casino ventures, 51–53
- West Virginia, 40
- West Virginia casino revenues, 42(*f*4.2)

"Casinos, Crime, and Community Costs" (Grinols & Mustard), 65

Casinos, economic/social effects of
- bankruptcy, 66
- casino occupations, 60*t*
- crime, 65
- Detroit, Michigan, casinos, contacts with minors in, 71(*t*6.4)
- disordered gambling, 66–68
- economics, 58–59
- employment/careers, 59–63
- job growth projections, gaming service occupations *vs.*. all occupations, 61*f*
- Louisiana riverboat casino employment data, 62(*t*6.3)
- Maryland Problem Gambling Helpline, callers to by gender, 69(*f*6.5)
- Maryland Problem Gambling Helpline, callers to by identity of caller, 70(*f*6.7)
- Maryland Problem Gambling Helpline, callers to by race/ethnicity, 70(*f*6.6)
- Maryland Problem Gambling Helpline, phone/text/online chat contacts received by, 68*f*
- Maryland Problem Gambling Helpline, problems reported by callers to, 69(*f*6.4)
- median annual wages, gaming service occupations *vs.* all occupations, 60*f*
- Missouri casinos, arrests at by month, 71(*t*6.5)
- national public opinion on, 57–58
- Nevada casino industry employment data, 62(*t*6.2)
- overview of, 57
- politics, 69–72
- suicide, 66
- tourism, 63–65
- underage gambling, 68

Casinos, Native American tribal
- California tribal casinos, 55–56
- Connecticut tribal casinos, 53–54
- Connecticut tribal gaming payments to state general fund, 54*t*
- economic effects of, 58
- employment in, effects on local economies, 61, 63
- federal labor laws and, 63
- federal recognition, 47–48
- gambling classes of, 46
- history of, 45–46
- overview of, 45
- political contributions of, 70
- regulation of, 46–47
- revenues of, 48–51
- tribal gaming revenues, by region, 50*t*
- tribal gaming revenues, growth in, 46*f*
- tribal gaming revenues, growth in, by region, 50*f*
- tribal gaming revenues, nationwide, 49*f*
- tribal-commercial casino ventures, 51–53

Casinos, online, 103

"Casinos and Political Corruption in the United States: A Granger Causality Analysis" (Walker & Calcagno), 69

"Casinos Use TV Stars to Draw New Customers" (Parry), 24

"Casinos/Gambling" (Center for Responsive Politics), 70

Center for Gaming Research, 9, 104–105

Center for Responsive Politics
- on casino industry political contributions, 70
- on monies spent on lobbying for casino industry, 71

"Changing Spousal Roles in and Their Effects on Recovery in Gamblers Anonymous: GamAnon, Social Support, Wives and Husbands" (Ferentzy, Skinner, & Antze), 19

Charities, gambling sponsored by, 12–13

Chen, Wai, 20

Cherokee tribal grouping, 45

Children. *See* Minors

Chippewa tribal grouping, 45

Christianity, 2

Christie, Chris, 93

Churchill Downs, 91–92

Citizens against Reservation Shopping, 52

Citizens against Reservation Shopping v. Haugrud, 52

Citizens against Reservation Shopping v. Zinke, 52

Civil Rights Act of 1964, 63

Clark County, Nevada, 26, 27

Class 1 gambling, 46

Class 2 gambling, 46

Class 3 gambling, 46

"The Clustering of Psychiatric Disorders in High-Risk Gambling Populations" (Abdollahnejad, Delfabbro, & Denson), 66

Coalition to Stop Internet Gambling, 102

Cobb, Ty, 99

Cockfighting, 99
"Cockfighting Fact Sheet" (Humane Society of the United States), 99
Cognitive behavior therapy, 19–20
Cognitive disorder, 85–86
College sports
  illegal gambling on, 98
  integrity of, 99–100
  wagering in Nevada on, 95
Colorado
  casino revenues of, 42(*t*4.23)
  commercial casinos of, 41
Commercial casinos. *See* Casinos, commercial
Compacts, online gaming, 106
Compacts, tribal-state. *See* Tribal-state compacts
"A Comparison of Individual and Group Cognitive-Behavioural Treatment for Female Pathological Gambling" (Dowling, Smith, & Thomas), 19
Comprehensive Annual Financial Report for the Years Ended June 30, 2016 and 2015 (Maryland Lottery and Gaming Control Agency), 43
Comps, casino, 24
Compulsive gambling, 17
  See also Gambling disorder; Pathological gamblers; Problem gamblers
Computer networks, lottery games and, 75
"Concerns Growing over How Spread of Online Gambling Will Play Out" (Hiltzik), 102
Connecticut
  tribal casinos, 53–54
  tribal gaming payments to state general fund, 54*t*
Connecticut Alliance against Casino Expansion, 53
Conner, Thaddieus, 51
Continental Congress, 2–3
Continental Wire Service, 98
Contracts, casino management, 51
Contrera, Jessica, 103
Convenience gambling, 10
Corporations
  casino owners/operators, 9–11
  gambling industry related, 11
  in Las Vegas gambling business, 4
Costello, Frank, 4
Cotti, Chad D., 61
Coursing, 92
Court cases
  California v. Cabazon Band of Mission Indians, 5, 46
  Citizens against Reservation Shopping v. Haugrud, 52
  Citizens against Reservation Shopping v. Zinke, 52
  Michigan v. Bay Mills Indian Community, 52
  Oneida Tribe of Indians v. State of Wisconsin, 45–46
  San Manuel Indian Bingo and Casino, 63
  Seminole Tribe of Florida v. Butterworth, 45
Covil, Wayne, 48
Craighill, Peyton
  on gamblers, 15–16
  on problem gamblers, 17
  on public opinion on gambling, 57
"A Crash Course in Vigorish . . . and It's Not 4.55%" (Martin), 94
CRDA (Casino Reinvestment Development Authority), 64–65
Crime
  arrests, by type of crime, 13*t*
  casinos and, 65
  illegal sports gambling, 97–99
  illegal sports gambling, effects on society, 99–100
  Internet gambling, effects of on, 107–108
  Mississippi, illegal gambling in, 33–34
  Missouri casinos, arrests at by month, 71(*t*6.5)
  online gambling, illegal, 106
  See also Illegal gambling; Illegal online gambling; Illegal sports gambling
Crime, organized
  in gaming industry, 12
  illegal sports gambling, link to, 98
  Las Vegas gambling and, 4
  in Nevada, 65
  in Nevada, eradication of, 25
  in Nevada, Senate hearings on, 94
Cruise ships, gambling on in Mississippi, 33
Cuomo, Andrew, 97
Cybernomads, 108

D

Daily Racing Form, 90
Daraban, Bogdan, 66
Darrow, Chuck, 24
Database of Criminal Animal Cruelty Cases (Pet-Abuse.com), 98–99
Davis, Gray, 55
Dayton, Mark, 77
"Dayton Vetoes Bill Dumping Instant-Play Online Lottery Games, Bar Ticket Sales at Gas Pumps" (Forliti), 77
Dead man's hand, 3
Deadwood, South Dakota
  casino revenues of, 43–44
  gambling in, history of, 43
Debate, supply *vs.* demand, 9
Delaware
  lottery tickets, online sales of, 103
  online gambling, legalization of in, 102
  online gaming financial data, 104*t*
  online poker compact with Nevada, 106
  sports gambling in, 93
Delfabbro, Paul, 66
Demirjian, Karoun, 106
Demographic Survey of Texas Lottery Players (Texas Lottery Commission), 78–79
Demographics
  of lottery players, 78–79
  of online gamblers, 106–107
  spending on Texas state lottery tickets by, 82*t*
  of sports betters, 98
  of Texas lottery players *vs.* nonplayers, 79*t*–80*t*
Denson, Linley, 66
Desai, Rani A., 16
Detroit, Michigan
  casino revenues, 40(*t*4.20)
  casino state taxes, 40(*t*4.21)
  casinos, contacts with minors, 71(*t*6.4)
  casinos/bankruptcy of, 40
"Detroit's Path into and out of Bankruptcy Is Paved with Casino Money" (Pierog & Lichterman), 40
Dexheimer, Eric, 84–85
Diagnostic and Statistical Manual of Mental Disorders (APA)
  on compulsive gambling, 6
  on criteria for gambling disorder, lottery players and, 85
  on pathological gambling, 17
  reclassification of gambling disorder, 18
Dice, 1
Disordered gambling
  casinos and, 66–67
  hotlines/treatment, 67–68
  self-exclusion programs, 67
"Disordered Gambling among University-Based Medical and Dental Patients: A Focus on Internet Gambling" (Ladd & Petry), 108–109
"Distribution of Registered US Foal Crop by State" (Jockey Club), 92
"Disulfiram, an Option for the Treatment of Pathological Gambling?" (Mutschler et al.), 20
Division of Special Revenue (Connecticut), 53
"Do Casinos Really Cause Crime?" (Walker), 65
Dodge City, Kansas, 43
"Does Legalized Gambling Elevate the Risk of Suicide?" (McCleary et al.), 66
Dogfighting
  arrests for, 98
  overview of, 99
Dogs. *See* Greyhound racing
DOJ. *See* U.S. Department of Justice

Dolan, Christopher B., 63
"Dot-cons," 108
Dougherty, Conor, 91–92
Dowling, Nicki, 19
DraftKings, 97
Dragonetti, Rosa, 19
Drugs, for treatment of pathological gambling, 20
Dunstan, Richard, 33

E

Easley, Mike, 75
Economic and Social Impact of Introducing Casino Gambling: A Review and Assessment of the Literature (Mallach)
on casinos, public opinion on, 6–7
on public opinion on gambling, 57
Economy
casinos, economic effects of, 58–59
division of lottery money, 81–82
government-regulated gambling, 13
Internet gambling, effects of on, 107
lotteries, economic effects of, 81
lottery funding, effects on education, 84–85
lottery income/apportionment of funds, by state, 74*t*
lottery winnings, taxes/withholding from, 84
lottery winnings, unclaimed, 82–84
Edelman, Marc, 97
Education
level of, lottery playing and, 78
lottery funding, effects of on, 84–85
vocational/professional for casino workers, 63
Edwards, Edwin, 35
"The Effect of Casinos on Local Labor Markets: A County Level Analysis" (Cotti), 61
"Effectiveness of Community-Based Treatment for Problem Gambling: A Quasi-experimental Evaluation of Cognitive-Behavioral vs. Twelve-Step Therapy" (Toneatto & Dragonetti), 19
Eisler, Kim Isaac, 53
El Rancho Vegas, 25
Electronic gaming
in casinos, 21–22
Illinois casino revenues from, 37
in Maryland, 42–43
at racinos, 5
small businesses and, 11
West Virginia casino revenues from, 41
"Elevated Suicide Levels Associated with Legalized Gambling" (Phillips, Welty, & Smith), 66
Elizabeth I, Queen of England
greyhound coursing and, 92
lottery established by, 2
"An Empirical Study of Gender Differences in Online Gambling" (McCormack, Shorter, & Griffiths), 107
"An Empirical Study of Personality Disorders among Treatment-Seeking Problem Gamblers" (Brown et al.), 20
Employment
casino occupations, 60*t*
in casinos, 59–63
casinos and, 58
job growth projections, gaming service occupations *vs.* all occupations, 61*f*
Louisiana riverboat casino, data on, 62(*t*6.3)
median annual wages, gaming service occupations *vs.* all occupations, 60*f*
Nevada casino industry, data on, 62(*t*6.2)
tribal casinos, federal labor laws and, 63
Encyclopaedia Britannica, 2
Environment, effects of on gamblers, 24
"Estimating the Effects of Casinos and of Lotteries on Bankruptcy: A Panel Data Set Approach" (Daraban & Thies), 66
Europa Star (cruise ship), 34
Evangelical Christians, 2
"Examining Gender Differences for Gambling Engagement and Gambling Problems among Emerging Adults" (Wong et al.), 16
Expanding the Vision: FY 2016 Annual Report (Maryland Center of Excellence on Problem Gambling), 67–68
"Experts: Online Gambling Slowed by Illegal Action" (AP), 106
"Extended Release Carbamazepine in the Treatment of Pathological Gambling: An Open-Label Study" (Black, Shaw, & Allen), 20

F

Facebook, casino games played on, 103
Fact Sheet: Greyhound Racing in the United States (Grey2K USA Worldwide), 93
"Facts at a Glance" (National Indian Gaming Commission), 5
Families, visiting Las Vegas, 64
FanDuel, 97
"FanDuel Secures an Important Victory in Daily Fantasy Sports Lawsuit, However Plaintiff Plans to Appeal" (Edelman), 97
"Fantasy Sports and Gambling: Line Is Blurred" (Brustein), 96
Fantasy Sports Trade Association, 96
Farkas, Karen, 86
Fay, Carly, 23
Federal Bureau of Investigation (FBI), 99
"Federal Employment Laws Impact Tribal Employers" (Allis), 63
Federal Insurance Contributions Act, 59
Federal Trade Commission (FTC), 109
Ferentzy, Peter, 19
Final Report (National Gambling Impact Study Commission), 6
Finnegan, Amanda, 64
Flamingo Hotel and Casino
Las Vegas opening of, 25
opening of, 4
Foals, Thoroughbred, 92
Fong, Timothy
on gambling addiction, states' role in, 85–86
on treatment of pathological gambling, 19
"For Schools, Lottery Payoffs Fall Short of Promises" (Stodghill & Nixon), 75
Forliti, Amy, 77
46th Annual Report of the California Horse Racing Board (California Horse Racing Board), 89
Foxwoods Casino and Resort, 53, 55
Franklin, Benjamin, 73
"Frequently Asked Questions" (North American Association of State and Provincial Lotteries), 77
Frontier gambling, 3
Frum, David, 58–59
FTC (Federal Trade Commission), 109
"FTC Warns Consumers about Online Gambling and Children" (FTC), 109
FullTiltPoker.com, 108
Funk, Cary
on gamblers, 15–16
on problem gamblers, 17
on public opinion on gambling, 57

G

Gainsbury, Sally M., 107
Gamblers
adults, 15–16
casino, overview of, 23–24
commercial casinos, consumer spending at, 14(*t*2.4)
Gamblers Anonymous 12-step recovery program, 19*t*
illegal sports, 97–98
lottery, demographics of, 78–79
lottery, group play, 80
lottery, overview of, 78
lottery, reasons for playing, 80–81
lottery winnings, taxes/withholding from, 84
19th-century Mississippi, 33
online, 106–107
pari-mutuel horse racing, takeout of, 89
problem, 17–20
questions for determining if a person is a compulsive gambler, 18*t*
young people, 16–17
Gamblers Anonymous
on compulsive gamblers, 17
establishment of, 6
gambling hotlines list of, 67

treatment provided by, 19
12-step recovery program of, illustrated, 19*t*
Gambler's fallacy, 23
Gambling
Americans who gambled in previous 12 months, by gambling type, 7*t*
casinos' methods for encouraging, 23–24
in England, 2
issues concerning/social impact of, 6
pari-mutuel gambling, 88–89
in precolonial/colonial U.S., 1–2
public opinion on, 6–7, 7*f*
in U.S., overview of, 1
in U.S. during 19th century, 2
in U.S. since 1900, 3–6
"Gambling, Gambling Activities, and Problem Gambling" (Holtgraves), 109
Gambling, game providers
arrests, by type of crime, estimated number of, 10*t*
casino gaming revenues, by state, 10*t*
casino owners/operators, 9–11
charities, 12–13
commercial casino gambling, consumer spending on, 14(*t*2.4)
criminals, 12
gambling corporations, 11
government, 13–15
Internet gambling businesses, 11–12
legal online operators, obstacles/growth potential for, 105–106
overview of, 9
small businesses, 11
tax revenues generated by casino gambling, by state, 15*t*
winnings reported to IRS, 14(*t*2.3)
Gambling, online. *See* Online gambling
Gambling addiction, 6
See also Pathological gamblers
"Gambling among Older, Primary-Care Patients" (Levens et al.), 16
Gambling: As the Take Rises, So Does Public Concern (Taylor, Funk, & Craighill)
on effect on community, 57
on gamblers, 15–16
on problem gamblers, 17
Gambling classes, Native American tribal casino, 46
Gambling disorder, 85–86
See also Pathological gamblers; Problem gamblers
Gambling Impact and Behavior Study: Report to the National Gambling Impact Study Commission (Gerstein et al.), 85
Gambling Impact Study (Spectrum Gaming Group)
on casino business model, advantages of, 58
on compacts for states online gambling, 103
on gambling activity, prevalence rates, 16
on monies wagered on legal/illegal sports betting, 12
on online gambling, broad customer base of, 107
on online poker, 101
on online wagering, potential of, 103
on public opinion on gambling, 58
Gambling in America (U.S. Commission on the Review of the National Policy toward Gambling), 4
Gambling in California (Dunstan), 33
Gambling in Connecticut: Analyzing the Economic and Social Impacts (Division of Special Revenue), 53
"Gambling in the South: Implications for Physicians" (Westphal et al.), 3
"Gambling Involvement and Increased Risk of Gambling Problems" (Phillips et al.), 109
"Gambling on College Sports: What's the Big Deal?" (NCAA), 99–100
"Gambling Severity, Impulsivity, and Psychopathology: Comparison of Treatment and Community-Recruited Pathological Gamblers" (Knezevic & Ledgerwood), 20
"Gambling with Our Future: City Poised to Hit Jackpot, or Lose Everything" (Kelly), 64–65
"GAME Act Proposing Repeal of Federal Prohibition on Sports Betting Revealed" (Purdum), 88
Game providers
online, 101–102
World Series of Poker, 101
Games
offered by casinos, 22–23
online, 103
sports, money and, 94
Games, online, 103
Gaming Accountability and Modernization Enhancement (GAME) Act, 88
Gaming machines, 22
Gaming Revenue Report: December 31, 2016 (Nevada Gaming Control Board), 26–27, 94
GAO (U.S. Government Accountability Office)
on legality of tribal casinos, 46
on tribal recognition process, 48
Garofalo, Pat, 84
"GCB Excluded Person List" (Nevada Gaming Commission & State Gaming Control Board), 12, 65
Georgia Lottery, online sales of, 103
Gerstein, Dean, 85
Gila River Indian Community, 71
Global Online Gambling (Marketline)
online gambling industry, growth projections for, 107
on online sports betting, 103
Global Payments, self-exclusion/self-limit services of, 67
Golden Pony Casino, 47
Goldstein, Sam, 20
"A Good Way to Wreck a Local Economy: Build Casinos" (Frum), 58–59
Gouker, Dustin, 102
Government
federal, gambling income of, 14
federal, gambling winnings reported to IRS, 14(*t*2.3)
federal, taxes on gambling, 14
gambling operations of, 13–14
state governments, gambling income of, 14–15
tribal governments, gambling income of, 15
Graham, Lindsey, 102
Grant, Jon E.
on opiate antagonists in treatment of pathological gambling, 20
"Pathologic Gambling and Bankruptcy," 66
Great Awakening, 2
Great Depression, 3
Great Recession
acceptability of gambling, effect on, 58
effects on Las Vegas tourism industry, 64
Greektown Casino, 40
Greer, Nancy, 23
Grey2K USA Worldwide, 93
Greyhound racing, 92–93
Griffiths, Mark D.
"An Empirical Study of Gender Differences in Online Gambling," 107
"Mom, Dad It's Only a Game! Perceived Gambling and Gaming Behaviors among Adolescents and Young Adults: An Exploratory Study," 109–110
"The Money Maze," 24
Grinols, Earl L., 65
"Gross Gaming Revenue Trending" (NIGC)
2015 tribal casino revenues, 58
2016 tribal casino revenues, 9, 21–22
"Gross Purses" (Jockey Club), 92
Gross wager, pari-mutuel gambling, 89
Group play, lottery, 80

H

Hancock, John, 73
Handbook of Resilience in Children (Goldstein & Brooks), 20
Handicapping, in horse racing, 91
Handle
in horse racing, 91
in pari-mutuel betting, 89
Harness racing, 90

Harrah's Entertainment, 6
Harris Poll, 87
Haugrud, Citizens against Reservation Shopping v., 52
Haugrud, Kevin, 52
"Health Correlates of Recreational Gambling in Older Adults" (Desai et al.), 16
Heitner, Darren, 58
"Here's How to Legally Gamble on the 2016 Race" (Contrera), 103
Hickok, Wild Bill, 3, 43
High rollers, 24
High-profit point tickets, 76
Hiltzik, Michael, 102
"Historical Lucky Retailers" (California State Lottery), 77
History
- of casinos, 21–22
- of gambling, 1–3
- of gambling in U.S., 3–6
- of Internet gambling, 101–103
- of lotteries, 73–75
- of Native American tribal casinos, 45–46

"History of Lotteries" (Louisiana Lottery Corporation), 73–74
"History of Playing-Cards" (International Playing-Card Society), 2
Hobson, Jeremy, 15
Hobson, Will, 88, 97
Hodgins, David C., 66
Holliday, Doc, 3
Holtgraves, Thomas, 109
Hoover Dam, 3
Horse racing
- economic effects of, 91–92
- during Great Depression, legalization of, 87
- horse races, betting on, 90–91
- in medieval times, 2
- non-Thoroughbred racing, 90
- North American Thoroughbred racing, total purse amounts, 92*f*
- North American Thoroughbred racing, total wager amounts, 91*t*
- overview of, 89–90
- pari-mutuel gambling and, 88–89
- pari-mutuel in Michigan, 39
- popularity of, 87
- Thoroughbred racetracks/races, 90

"Horse Racing's Slide Spurs New Bet on Track Land" (Dougherty), 91–92
Hot Lotto, 77
Hotlines, disordered gambling, 67–68
House advantage
- of casinos, by game, 23*t*
- overview of, 23
- statistical, 1

"The House Advantage: A Guide to Understanding the Odds" (AGA)
- house advantage, variation of by region/casino, 23
- on statistical advantage of the house, 1

"How Much Sports Gambling Is Going On out There?" (Keating), 97–98
"How the Internet Is Changing Gambling: Findings from an Australian Prevalence Survey" (Gainsbury et al.), 107
Howard, Theresa, 64
Hubbub, 2
Hughes, Howard, 4
Humane Society of the United States, 99

I

Ideation, suicidal, 66
IGRA. *See* Indian Gaming Regulatory Act
Illegal gambling, 21st century, 5
Illegal online gambling, 107
Illegal sports gambling
- in college basketball, 87
- Nevada sports book, link to, 98
- organized crime, link to, 98
- overview of, 97–98

Illinois
- commercial casinos of, 36–37
- gaming revenues, 38(*t*4.16)
- gaming revenues, by casino/type, 38(*t*4.17)
- online lottery sales of, 77
- self-exclusion program of, 67
- state *vs.* local share of gaming taxes, 39(*t*4.18)

Illinois Loss Recovery Act, 97
Illinois Lottery
- online sales of, 103
- scratch game of, 76

Increasing the Odds: Volume 3, Gambling and the Public Health, Part 1 (National Center for Responsible Gaming), 67
"Indian Casinos Have Different Set of Laws" (Dolan), 63
"Indian Entities Recognized and Eligible to Receive Services from the United States Bureau of Indian Affairs" (BIA), 47
Indian Gaming Regulatory Act (IGRA)
- description of, 5
- distribution of tribal casino earnings, 51
- passage of, 47

Indian Issues: Improvements Needed in Tribal Recognition Process (GAO), 48
Indian Issues: Timeliness of the Tribal Recognition Process Has Improved, but It Will Take Years to Clear the Existing Backlog of Petitions (GAO), 48
Indian Reorganization Act, 48
Indiana
- casino revenues and taxes, 33*t*
- commercial casinos of, 33

Indiana Council on Problem Gambling (ICPG), 67
Indiana Gaming Commission, 33
Indianz.com, 52
"Industry Demographics" (Fantasy Sports Trade Association), 96
Internal Revenue Service (IRS)
- gambling winnings reported to, 14(*t*2.3)
- lottery winnings reported to, 84

International Playing-Card Society, 2
Internet gambling
- Delaware online gaming financial data, 104*t*
- effects of, 107–110
- history/legal status of, 101–103
- New Jersey online gambling revenues, 104*f*
- online gamblers, 106–107
- online games, 103
- overview of, 101
- revenue levels/projections, 103–106
- *See also* Online gambling

"Internet Gambling: An Emerging Concern in Family Practice Medicine" (Petry), 109
"Internet Gambling Grows among Male Youth Ages 18 to 22" (Annenberg Public Policy Center), 16, 109
Internet Gambling Prohibition Act, 71
Iowa
- casino financial data, 41*t*
- commercial casinos of, 40–41

Iowa Racing and Gaming Commission, 41
"Is the House Winning? Exploring the Impact of Indian Gaming" (Conner & Taggart), 51

J

Jackpot, lottery, 76
Jackpot Captain program, 80
Jackpot fatigue, 86
Jackson, Andrew, 3
Jacksonian era, 3
Jai alai, 89
James I, King of England, 2, 73
Jefferson, Thomas, 89
Jensen, Katherine, 43
Jobs
- gaming service, Nevada casino, 61
- projected growth rate for gaming industry, 60–61
- *See also* Employment

Jockey Club
- establishment of, 89
- on gross purse of Thoroughbred racing, 92
- on horse racing in U.S., early history of, 89
- on number of horse races run annually, 90
- on pari-mutuel handle from Thoroughbred horse racing, 91

Johnson, Cuthbert William, 2
Jones, Jeffrey M., 6–7
Journal of Gambling Studies (National Council on Problem Gambling), 19

K

Kagan, Elena, 52
Kansas
commercial casinos of, 43
state-owned casinos of, 15
Kansas Racing and Gaming Commission, 43
Kasler, Dale, 51
Keating, Peter, 97–98
Kefauver, Estes, 94
Kelly, Mike, 64–65
Keno, 75
Kentucky, 91
Kentucky Derby, 90
Kerzner, Sol, 53
King's Club, 46
Knezevic, Bojana, 20
Koch, Ed, 4

L

Labor laws, tribal casinos and, 63
Ladd, George T., 108–109
"Landless Tribe Waits Federal Recognition" (AP), 48
Lansky, Meyer, 4
Las Vegas, Nevada
commercial casinos of, 26–28
downtown gaming revenue, by type, 29*t*
gambling halls in, early history of, 4
strip gaming revenue, by type, 28*t*
tourism in, 64
Las Vegas Convention and Visitors Authority (LVCVA)
on Nevada gambling revenues, 25
on tourism in Las Vegas, 64
visitors to Las Vegas, profile of, 28
Las Vegas Sands Corporation, 70
Las Vegas Sports Consultants, Inc., 94
Las Vegas Visitor Profile Study: 2016 (LVCVA), 28
The Last Gamble: Betting on the Future in Four Rocky Mountain Mining Towns (Jensen Blevins), 43
"Latest Gaming Industry Report: Indian Gaming Made Small Gains in 2011" (Toensing), 60
Law Enforcement Summit on Illegal Sports Betting: After-Action Report (AGA), 97
The Law of Bills of Exchange, Promissory Notes, Checks, &c (Johnson), 2
Lazarus, David, 15
Ledgerwood, David M., 20
Legal gambling, history of, 12
Legal sports gambling
bookmaking, 94
fantasy sports, 96–97
low-stakes, 95–96
money/games, 95
in Nevada, developments in, 94
overview of, 93–94
sports/race books in Nevada, 94
Legal status, Internet gambling, 101–103
"Legalize and Regulate Sports Betting" (Silver), 88
Legislation and international treaties
Americans with Disabilities Act, 63
Articles of Association, 2–3
Civil Rights Act of 1964, 63
Federal Insurance Contributions Act, 59
Gaming Accountability and Modernization Enhancement (GAME) Act, 88
Illinois Loss Recovery Act, 97
Indian Gaming Regulatory Act, 5, 47, 51
Indian Reorganization Act, 48
Internet Gambling Prohibition Act, 71
National Labor Relations Act, 63
Pennsylvania Race Horse Development and Gaming Act, 31–32
Professional and Amateur Sports Protection Act, 87, 93
Racetrack Table Games Act (West Virginia), 41
Racketeer Influenced and Corrupt Organizations Act, 12, 25
Unlawful Internet Gambling Enforcement Act, 5, 11, 96
Wire Act, 5, 12, 102, 107
Lesieur, Henry R., 17
Levens, Suzi, 16
Lewis, Danny, 2
Lichterman, Joseph, 40
Lighting, 24
List of Excluded Persons (Nevada Gaming Commission/State Gaming Control Board), 12
Liu, Yixin, 14
Loaded dice, 1
Lobbying, 69–71
Long, Earl Kemp, 35
Lots, drawing of, 73
Lotteries
California government finances, 85*t*
economic/social effects of, 81–86
future of, 86
history of, 73–75
lottery contributions to beneficiaries, by state, 83*t*–84*t*
lottery games, 75–77
lottery income/apportionment of funds, by state, 74*t*
lottery operations, overview of, 77–78
median spending on Texas state lottery tickets, by demographic characteristics of players, 82*t*
of medieval England, 2
19th century, 3
online sales of, 103
overview of, 73
players of, 78–81
in precolonial/colonial U.S., 2
state-sponsored, 5
Texas lottery game, percentage of people who played, 80*f*
Texas lottery players *vs.* nonplayers, demographic characteristics of, 79*t*–80*t*
"Lottery Gambling: A Review" (Ariyabuddhiphongs)
on lotteries, at-risk gamblers and, 85
on lotteries, uniqueness of, 76
on lottery players, demographics of, 78
on playing lottery, reasons for, 80–81
Lottery games
overview of, 75–76
Powerball/Mega Millions, 76–77
scratch games, 76
Lottery ticket sales, online, and Wire Act, 5, 12
Lotto captains, group play, 80
Lotto games, 75
Louis IX, King of France, 2
Louisiana
commercial casinos of, 35–36
land-based casino financial data, 37(*t*4.14)
19th-century gambling in, 3
political corruption in, 71
riverboat casino employment, data on, 62(*t*6.3)
riverboat gaming statistics, 36*t*
video gaming statistics, 38(*t*4.15)
Louisiana Gaming Control Board, 36
Louisiana Lottery Company, 73–74
Louisiana Lottery Corporation, 73–74
Lovett, Ian, 52
Low-stakes sports gambling, 95–96
Luck, 23
LVCVA. *See* Las Vegas Convention and Visitors Authority

M

Macau, 26
"Macau's 2013 Gambling Revenue Rose 19% to $45.2 Billion" (O'Keeffe), 26
Mahabharata, 1
Maine, 43
Maine Gambling Control Board, 43
Make Me Rich! (television show), 76
Malcolm, John G., 108
Mallach, Alan
on casinos, public opinion on, 6–7
on public opinion on gambling, 57
Manning, Mary, 4
Mansley, Chrystal, 66
March Madness, 100

"March Madness and American Gambling Habits" (Hobson), 15
Markell, Jack, 106
Marketline
on online gambling, growth of, 103
online gambling industry, growth projections for, 107
on online sports betting, 103
Martin, J. R., 94
Maryland
commercial casinos of, 42–43
horse breeding program of, 92
Maryland Center of Excellence on Problem Gambling, 67–68
Maryland Jockey Club, 89
Maryland Lottery and Gaming Control Agency, 43
Maryland Problem Gambling Helpline
callers to by gender, 69(*f*6.5)
callers to by identity of caller, 70(*f*6.7)
callers to by race/ethnicity, 70(*f*6.6)
phone/text/online chat contacts received, 68*f*
problems reported by callers, 69(*f*6.4)
Mashantucket Pequot Tribe, 53
Massachusetts, 76
Masterson, Bat, 3
McCleary, Richard, 66
McCormack, Abby, 107
McMullan, John L., 108
McNeil, Charles K., 94
Mechanical reel machines, 22
Medieval times, 2
"Mega Failure: Why Lotteries Are a Bad Bet for State Budgets" (Garofalo), 84
Mega Millions, 76–77
"Mega Millions: Do Lotteries Really Benefit Public Schools?" (Strauss), 84
MegaHits (video lottery game), 77
Melia, Michael, 48
Men, 68
Menominee Tribe, 52
Mexican American tribal grouping, 45
MGM Grand
Las Vegas ad campaign, 64
in Michigan, 39–40
MGM Resorts International
noncasino revenues of, 27
revenues generated by, 10
Michigan
Bay Mills Indian Community off-reservation casino, opposition to, 52
commercial casinos of, 39–40
Detroit, casino revenues of, 40(*t*4.20)
Detroit, casino state taxes paid by, 40(*t*4.21)
self-exclusion program of, 67
Michigan Gaming Control Board
on Michigan casino revenues, 40
on minors attempting to gamble in casinos, 68
Michigan Lottery, 76
Michigan v. Bay Mills Indian Community, 52
Middle ages, 2
Miller, Donald E., 85
Minnesota Lottery
banning of online lottery sales, 77
lottery tickets, online sales of, 103
Minors
casinos contacts with, 71(*t*6.4)
Internet gambling, underage, 109–110
underage gambling among, 68
See also Young people
Mississippi
casino revenues, by region, 34*t*
casino tables hold, by game and region, 35*t*
commercial casinos of, 33–35
political corruption, address to, 71
Tunica County, economic effects of casinos of, 61
Mississippi River, 3
Missouri
casino gaming financial data, by casino location, 39(*t*4.19)
casinos, arrests at by month, 71(*t*6.5)
commercial casinos of, 37–39
online lottery sales of, 77
Voluntary Exclusion Program of, 67
Missouri Gaming Commission, 68
Mizzi, Sorel "Imper1um," 108
"Mob Ties" (Koch & Manning), 4
Mohegan Sun, 53
Mohegan Tribe, 53
"Mom, Dad It's Only a Game! Perceived Gambling and Gaming Behaviors among Adolescents and Young Adults: An Exploratory Study" (Calado, Alexandre, & Griffiths), 109–110
Money. *See* Revenues, gambling
Money laundering, 107–108
"The Money Maze" (Griffiths), 24
Moneymaker, Chris, 101
Mono Indians, 52
Monshimout, 2
Montana, 95–96
Mothers against Drunk Driving, 53
Motion Picture Association of America (MPAA)
on film industries 2015 revenues, 58
on U.S. movie industry revenues, 9
MotorCity Casino, 39–40
Muckleshoot Tribe, 51
"Multicenter Investigation of the Opioid Antagonist Nalmefenein the Treatment of Pathological Gambling" (Grant et al.), 20
Multi-State Lottery Association (MUSL), 76
MUSL (Multi-State Lottery Association), 76
Mustard, David B., 65
Mutschler, Jochen, 20

N

NAFTM 2015 Annual Report (National Association of Fundraising Ticket Manufacturers), 13
NAICS (North American Industry Classification System), 21
National Academies Press
on pathological gamblers, numbers of, 18
on pathological gambling, 6
National Annenberg Survey of Youth (Annenberg Public Policy Center)
on Internet gambling among youth, 109
on young people, gambling among, 16
National Association of Fundraising Ticket Manufacturers, 13
National Association of Trotting Horse Breeders, 90
National Center for Responsible Gaming, 67
National Certified Gambling Counselor program, 19
National Collegiate Athletic Association (NCAA)
on March Madness tournament, monies wagered on, 99–100
on student athletes involved in sports gambling, 100
National Council on Problem Gambling (NCPG)
National Problem Gambling Helpline Network of, 67
"NCPG Affiliate Member List," 19
National Football League (NFL), 93
National Gambling Impact Study Commission, 6
National Greyhound Association (NGA), 93
National Incident-Based Reporting System—Volume 1: Data Collection Guidelines (FBI), 99
National Indian Gaming Commission (NIGC)
prosecution of Seminole Tribe, 47
role of, 45, 46–47
on small casino earnings, 50
states with tribal gaming, number of, 5
on tribal casinos, revenues generated by, 9, 21–22, 58
tribal revenues, oversight of, 48–50
on tribes with casino management contracts, 51
National Labor Relations Act (NLRA), 63
National Labor Relations Board (NLRB), 63
National Problem Gambling Helpline Network (National Council on Problem Gambling), 67

Native Americans and Alaska Natives
bingo halls/casinos operated by, 5
colonial era, gambling among, 2
legal victories of in support of native lands casinos, 21
See also Casinos, Native American tribal
"Native Americans Can't Always Cash In on Casinos" (Wells), 51
Navajo Nation tribal grouping, 45
NCAA. *See* National Collegiate Athletic Association
NCAA Student-Athlete Gambling Behaviors and Attitudes: 2004–2012 (NCAA), 100
NCPG. *See* National Council on Problem Gambling
"NCPG Affiliate Member List" (National Council on Problem Gambling), 19
Nelson, Sara E., 67
Networks, computer, 75
Nevada
casino employees, numbers of, 59–60
commercial casinos of, 25–28
employment data on casino industry, 62(*t*6.2)
gaming revenue of, 26*f*
gaming revenue of, by type, 27*t*
Las Vegas downtown gaming revenue, by type, 29*t*
Las Vegas strip gaming revenue, by type, 28*t*
legalized casinos in, 3–4
online gambling, legalization of in, 102
online poker compact with Delaware, 106
sports betting, amount wagered on by sport, 95*t*
sports betting financial data, 88*t*
sports betting revenue percentages, by sport, 97*t*
sports betting revenues, by sport, 96*t*
sports gambling in, 93
sports/race books in, 94
"Nevada, Delaware Ink Online Poker Pact to Expand Player, Revenue Pools" (Demirjian), 106
Nevada Gaming Commission
establishment of, 4
List of Excluded Persons of, 12
and State Gaming Control Board, 26, 65
Nevada Gaming Control Board
on fantasy sports, gaming licenses for, 97
on sports pools/race books operating in Nevada, 94
Nevada sports book, 98
Nevada State Gaming Control Board, 26
Nevada Statewide Casino Employment—Productivity, Revenues, and Payrolls: A Statistical Study, 1990–2016 (Schwartz & Rajnoor), 61
Nevada's Black Book, 12
"New Bill Would Prohibit Internet Gambling, Including Where Already Legal" (Tetreault), 102
New Hampshire
education funding sweepstakes of, 75
state-sponsored lottery of, 5
New Hampshire Lottery, 5, 75
New Jersey
Atlantic City casino industry employment figures, by casino, 32(*t*4.7)
Atlantic City casino industry revenues, by source of revenue, 30(*t*4.5)
casino gaming revenue, 30(*t*4.4)
casino gaming revenue, by type, 31*t*
commercial casinos of, 28–29
online gambling, legalization of in, 102
online gambling revenues, 104*f*
self-exclusion program of, 67
sports betting in, court action on, 93–94
New Jersey Casino Control Commission, 65
"New Live-Entertainment Venues Energize Casinos and Their Patrons" (Darrow), 24
New Orleans, Louisiana
casinos of, 36
19th-century gambling in, 3
"New Report Shows Strong Ties between Illegal Gambling and Organized Crime" (AGA), 5
New York
FanDuel/DraftKings ban in, 97
state-sponsored lottery of, 5
New York Lottery, 5, 75
News Media Alliance, 98
NFL (National Football League), 93
NGA (National Greyhound Association), 93
NIGC. *See* National Indian Gaming Commission
19th century, gambling during, 3
Nixon, Ron, 75
NLRA (National Labor Relations Act), 63
NLRB (National Labor Relations Board), 63
Nonbanked games, 22
Noncasino revenues, 27–28
Nongaming revenues, 58
Non-Thoroughbred horse racing, 90
North American Association of State and Provincial Lotteries
"Frequently Asked Questions," 77
on state/local revenues from gambling, 15
North American Gaming Almanac (Casino City Press), 10
North American Industry Classification System (NAICS), 21
"North American Sports Market at $75.7 Billion by 2020, Led by Media Rights" (Heitner), 58
North American Thoroughbred racing
total purse amounts, 92*f*
total wager amounts, 91*t*
North Fork Band of Mono Indians, 52
"Number of Races" (Jockey Club), 90
Numbers games
description of, 75
organized crime involvement in, 12

O

Obama, Barack
federal stimulus funds, stance on, 64
off-reservation casinos, administration's attitude towards, 52
Odds
against gamblers in casinos, 23
in horse race betting, 90
Offtrack betting (OTB)
facilities for pari-mutuel wagering, 89
in Nevada, legalization of, 94
Ogle, Samuel, 89
Ohio
acceptability of gambling in, 58
commercial casinos of, 41–42
Ohio Casino Control Commission, 42
"Ohio Lottery Considering Offering a $50 Instant Ticket" (Farkas), 86
O'Keeffe, Kate
"Macau's 2013 Gambling Revenue Rose 19% to $45.2 Billion," 26
"Online Gambling Suffers Setback," 103
Oller, Pierre, 88
"One Decade of Self Exclusion: Missouri Casino Self-Excluders Four to Ten Years after Enrollment" (Nelson et al.), 67
Oneida Indian Nation
bingo games of, 45
political contributions of, 70
Oneida Tribe of Indians v. State of Wisconsin, 45
"Online Crime and Internet Gambling" (McMullan & Rege), 108
Online gambling
controversy over, 10
illegal sports, 98
Internet gambling, 6–7
Internet gambling businesses, 11–12
lottery ticket sales, 77–78
sports betting, 103
sports betting, illegality of, 107
sports gambling, 12
See also Internet gambling
"Online Gambling and Poker Bill Tracker" (Gouker), 102
Online Gambling Five Years after UIGEA (Stewart), 106
"Online Gambling Suffers Setback" (O'Keeffe), 103
Online lottery sales, allowance, for Wire Act, 102
Operation Bet Smart campaign, 6
Operations, lottery, 77–78

Opiate antagonists, 20
"Opiate Antagonists in Treatment of Pathological Gambling" (Grant et al.), 20
"An Overview of and Rationale for Changes Proposed for Pathological Gambling in DSM-5" (Petry et al.), 18–19

P

"Pamunkey Indians Become First Virginia Tribe Given Federal Recognition" (Covil), 48
Pamunkey Tribe, 48
Pari-mutuel gambling
  dog racing, states permitting, 93
  horse racing in Michigan, 39
  overview of, 88–89
"Pari-Mutuel Handle" (Jockey Club), 91
Parlay bet, 93
Parry, Wayne, 24
Passive drawing games, 75
"Pathologic Gambling and Bankruptcy" (Grant et al.), 66
Pathological gamblers
  description of, 17
  Maryland Problem Gambling Helpline, callers to by gender, 69(*f*6.5)
  Maryland Problem Gambling Helpline, callers to by identity of caller, 70(*f*6.7)
  Maryland Problem Gambling Helpline, callers to by race/ethnicity, 70(*f*6.6)
  Maryland Problem Gambling Helpline, phone/text/online chat contacts received by, 68*f*
  Maryland Problem Gambling Helpline, problems reported by callers to, 69(*f*6.4)
  self-exclusion programs for, 67
  in U.S., 6
  See also Disordered gambling; Problem gamblers
Pathological Gambling: A Critical Review (National Academies Press)
  on gambling addiction, 6
  overview of, 18
"Pathological Gambling: Update on Assessment and Treatment" (Fong), 19
Pennsylvania
  casino revenues of, 32(*t*4.8)
  casinos of, effects on Atlantic City's casinos, 29
  commercial casinos of, 31–32
Pennsylvania Gaming Control Board, 31
"Pennsylvania Leads U.S. in Gaming with Tax Windfall" (Varghese), 32
Pennsylvania Race Horse Development and Gaming Act, 31–32
People for the Ethical Treatment of Animals, 93
Perks, casino, 24
Pet-Abuse.com, 98
Petry, Nancy M.
  "Disordered Gambling among University-Based Medical and Dental Patients: A Focus on Internet Gambling," 108–109
  "Internet Gambling: An Emerging Concern in Family Practice Medicine," 109
  "An Overview of and Rationale for Changes Proposed for Pathological Gambling in DSM-5," 18–19
Pettorruso, Mauro, 20
Philadelphia, Pennsylvania, 32
Phillips, David P., 66
Phillips, James G., 109
Pierog, Karen, 40
Players. *See* Gamblers
Playing cards, 2
Poker
  banked/nonbanked, descriptions of, 22
  in 19th-century U.S., 3
  online, banning of, 11–12
  online, effects of UIGEA on, 102
  World Series of Poker, effects of on online poker, 101
Poker Alice, 43
Politics, casinos and, 69–72
Poverty rates, lottery sales and, 78
Powerball, 76–77
PredictIt.org, 103
PredictWise.com, 103
"Pro Football Is Still America's Favorite Sport" (Harris Poll), 87
Problem gamblers
  gambling disorder, 18–19
  Internet gambling, effects of on, 107–108
  overview of, 17–18
  treatment methods, 19–20
  treatment organizations, 19
  See also Pathological gamblers
Professional and Amateur Sports Protection Act, 87, 93
Prohibition Era, 4
Promises Made, Promises Broken: An Overview of Gambling in New Jersey and Recommendations for the Future (Liu), 14
"A Psychological Autopsy Study of Pathological Gamblers Who Died by Suicide" (Wong et al.), 66
Psychology in casino design, 24
Public opinion
  casinos, acceptability of, 22
  on gambling, 6–7
  on gambling in 1970s, 4
  national on casinos, 57–58
  on selected issues, 7*f*
Pull tabs, 75
Purdum, David, 88
Purse, horse racing, 92

Q

Quarter horse racing, 90
"Queen Elizabeth I Held England's First Official Lottery 450 Years Ago" (Lewis), 2
"Questions & Answers about Gamblers Anonymous" (Gamblers Anonymous), 17–18
Quick Pick, 75

R

"Race and Track Information" (American Quarter Horse Association), 90
Race book
  description of, 89
  in Nevada, 94
"Race Meet Calendar 2017" (Arabian Jockey Club), 90
Racetrack Table Games Act (West Virginia), 41
"Racing Links: Race Tracks" (Daily Racing Form), 90
Racinos
  description of, 90
  racetracks with, competitive advantage of, 91
Racketeer Influenced and Corrupt Organizations Act
  effects on casino business, 25
  purpose of, 12
Rajnoor, Alexis, 61
Rancherias, 55
Rebuck, David, 106
Reconstruction, 73–74
"Red Hawk Casino's Fortunes Have Disappointed So Far" (Kasler), 51
Referendum on Indian Gaming Compacts (California), 52
Rege, Aunshul, 108
Regressive tax, 78, 81
Regulation, Native American tribal casino, 46–47
Regulation 14 (Nevada Gaming Commission), 68
Regulatory capture, 69–70
Reno, Nevada, 3
Reservation shopping, 52
Reservations, Native American, 51–53
"Resilience and Self-Control Impairment" (Chen & Taylor), 20
Retailers, lottery, 77–78
Revenge of the Pequots: How a Small Native American Tribe Created the World's Most Profitable Casino (Eisler), 53
"Revenue Totals" (Maine Gambling Control Board), 43
Revenues, gambling
  Atlantic City casino industry, by source of revenue, 30(*t*4.5)
  of California horse racetracks, 89

casino, 21
casino gaming revenues, by state, 10*t*
Colorado casino, 41, 42(*t*4.23)
Detroit, Michigan, casino, 40(*t*4.20)
economic effects of, 58–59
gambling winnings reported to IRS, 14(*t*2.3)
government, 59
illegal online in California, 106
Illinois casino, 37
Illinois gaming, 38(*t*4.16)
Illinois gaming, by casino/type, 38(*t*4.17)
Indiana casino, 33, 33*t*
Internet gambling levels/projections, 103–106
Iowa casino, 41
Iowa casino financial data, 41*t*
Las Vegas downtown gaming revenue, by type, 29*t*
Las Vegas strip gaming revenue, by type, 28*t*
lottery, division of, 81–82
lottery, pressure for increased, 86
Louisiana casino, 36
Louisiana land-based casino financial data, 37(*t*4.14)
Louisiana riverboat gaming statistics of, 36*t*
Louisiana video gaming statistics, 38(*t*4.15)
Maine casino, 43
Maryland casino, 43
Michigan casino, 40
Mississippi casino, 34–35
Mississippi casino, by region, 34*t*
Missouri casino, 37
Missouri casino gaming financial data, by casino location, 39(*t*4.19)
of Native American tribal casinos, 48–51
Nevada casino, 25–27
Nevada gaming, 26*f*
Nevada gaming, by type, 27*t*
New Jersey casino, 29
New Jersey casino gaming, 30(*t*4.4)
New Jersey casino gaming, by type, 31*t*
online gambling, 103, 104, 107
Pennsylvania casino, 32, 32(*t*4.8)
riverboat casino financial data, by state, 37(*t*4.13)
South Dakota casino, 44*f*
sports gambling, 94
of state lotteries, 73
tax revenues generated by casino gambling, by state, 15*t*
tribal casinos, 45
tribal casinos in Connecticut, 53, 55
tribal gaming, by region, 50*t*
tribal gaming, growth in, 46*f*
tribal gaming, growth in, by region, 50*f*
tribal gaming, nationwide, 49*f*
West Virginia casino, 41, 42(*f*4.2)
See also Economy
Rincon Band of Luiseño Indians, 55
"Rincon Tribe Wins Slot Suit against State" (Soto), 55–56
"The Rise and Fall of the Third Wave: Gambling Will Be Outlawed in Forty Years" (Rose), 3
"Risk Factors for Suicide Ideation and Attempts among Pathological Gamblers" (Hodgins, Mansley, & Thygesen), 66
Riverboats
casino financial data, by state, 37(*t*4.13)
Illinois casino, 36–37
Indiana casino, 33
Iowa casino, 41
Louisiana casino, 36, 36*t*
Missouri casino, 37–39
19th century, gambling on, 3
Rockloff, Matthew J., 23
Rome, ancient, 87
Romer, Daniel, 16, 109
Rose, I. Nelson, 3
Rose, Pete, 99
Rosenthal, Frank "Lefty," 94

S

Sacramento region, tribal gaming revenues of, 55
Salaries, casino employee, 60
Saloons, 3
San Manuel Indian Bingo and Casino (court case), 63
Saratoga racecourse, 89
Schneiderman, Eric T., 97
School of Continuing Studies (Tulane University), 63
"Schools Lose Out in Lotteries" (Miller), 85
Schuetz, Richard, 106
Schwartz, David G., 61
Schwarzenegger, Arnold, 55
Scratch games, 76
Self-exclusion programs
overview of, 67
for pathological gamblers, 67
Seminole Tribe
bingo games of, 45
gaming violations of, 47
"Seminole Tribe Accused of Violating Federal Laws" (Weaver), 47
Seminole Tribe of Florida v. Butterworth, 45
Senior citizens, 16
Shaw, Martha C., 20
Shorter, Gillian W., 107
Showboat Star (riverboat), 36
Siegel, Bugsy, 4, 25
"Signs of a Surge in Las Vegas Conventions" (Velotta), 64
Silver, Adam, 88
Simon, Stephanie, 43
Simulcasting, 89
Sioux tribal grouping, 45
Skinner, Wayne, 19
Slot machines
casino, popularity of, 22
modern, 22–23
online, 103
at racetracks, 90, 91
"Small Business Size Standards: Arts, Entertainment, and Recreation" (SBA), 50–51
Small businesses, 11
Smith, David, 19
Smith, Marisa M., 66
Smith, Owen Patrick, 93
Snyder, Jimmy, 94
"The Social Contagion of Gambling: How Venue Size Contributes to Player Losses" (Rockloff, Greer, & Fay), 23
Social impact
of gambling, 6
of illegal sports gambling, 99–100
of lotteries, compulsive gambling/cognitive disorder and, 85–86
of lotteries, overview of, 81
lottery players, socioeconomic status of, 78
Social networking, 103
SOGS. *See* South Oaks Gambling Screen
Soto, Onell R., 55–56
South Dakota
casino revenues of, 43–44, 44*f*
commercial casinos of, 43
South Dakota Commission on Gaming, 43
South Oaks Gambling Screen (SOGS)
description of, 17
Internet gamblers scores on, 108
Speaking Rock Entertainment Center, 72
Spectrum Gaming Group
on casino business model, advantages of, 58
on compacts for states online gambling, 103
on gambling prevalence rates, 16
on monies wagered on legal/illegal sports betting, 12
on online gambling, broad customer base of, 107
on online poker, 101
on online wagering, potential of, 103
on public opinion on gambling, 58
Spiel, 75
Spokane Tribe, 52
Sport of kings. *See* Horse racing
Sport of queens. *See* Greyhound racing
Sports book, 94

Sports gambling
animal fighting, 98–99
bookmaking and, 94
greyhound racing, 92–93
horse racing, 89–92
illegal, 97–98
illegal, effects on society, 99–100
legal, 93–97
in Nevada, developments in, 94
Nevada sports betting, amount wagered on by sport, 95*t*
Nevada sports betting financial data, 88*t*
Nevada sports betting revenue percentages, by sport, 97*t*
Nevada sports betting revenues, by sport, 96*t*
North American Thoroughbred racing, total purse amounts, 92*f*
North American Thoroughbred racing, total wager amounts, 91*t*
online, illegality of, 12
overview of, 87
pari-mutuel gambling, 88–89
prevalence of, 87–88
sports/race books in Nevada, 94
"Sports Gambling in U.S.: Too Prevalent to Remain Illegal?" (Hobson), 88, 97
Sports tab game, 95–96
Standardbred horses, 90
"State by State" (Grey2K USA Worldwide), 93
"State Funded Help Line Calls and State Funded Problem Gambling Treatment Enrollment Data by Indiana County by Year" (ICPG), 67
State Gaming Control Board (Nevada), 12
"(State) House Rules in Kansas Casino" (Simon), 43
State of Wisconsin, Oneida Tribe of Indians v., 45–46
States
casino gaming revenues by state, 10*t*
charitable gambling regulated by, 12–13
with commercial casinos operating in, 22
Connecticut tribal gaming payments to state general fund, 54*t*
with dog racing bans, 93
establishment of lotteries, 75
gambling hotlines of, 67
gambling in, 5
government revenues from gambling, 59
with legal Internet gambling, 101
with legalized gambling, mid-1990s, 4
lotteries operated by, 15
lottery contributions to beneficiaries, by state, 83*t*–84*t*
lottery income/apportionment of funds, by state, 74*t*
lottery operations by, 73
online gaming compacts, multistate, 106
permitting low-stakes sports gambling, 95–96
Powerball/Mega Millions games of, 76–77
scratch games of, 76
state-sponsored lotteries, 5
tax revenues from gambling, 59
tax revenues generated by casino gambling, by state, 15*t*
taxes on gambling, 14–15
tribal gaming, role in regulation of, 47
without lotteries, 86
States, legal Internet gambling, in Wire Act, 107
"The States' Role in Gambling Addiction" (Fong), 85–86
Station Casinos
political contributions of, 70
revenues generated by, 10
Thunder Valley Casino, management of, 51
Statistical information
Americans who gambled in previous 12 months, by gambling type, 7*t*
arrests, by type of crime, estimated number of, 13*t*
Atlantic City casino industry employment figures, by casino, 32(*t*4.7)
Atlantic City casino industry revenues, by source of revenue, 30(*t*4.5)
California government finances, 85*t*
casino gaming revenues, by state, 10*t*
casinos, house advantage of, by game, 23*t*
Colorado casino revenues, 42(*t*4.23)
commercial casino gambling, consumer spending on, 14(*t*2.4)
Connecticut tribal gaming payments to state general fund, 54*t*
Delaware online gaming financial data, 104*t*
Detroit, Michigan, casino revenues, 40(*t*4.20)
Detroit, Michigan, casino state taxes, 40(*t*4.21)
Detroit, Michigan, casinos, contacts with minors in, 71(*t*6.4)
Illinois gaming revenues, 38(*t*4.16)
Illinois gaming revenues, by casino/type, 38(*t*4.17)
Illinois gaming taxes, state *vs.* local share of, 39(*t*4.18)
Indiana casino revenues/taxes, 33*t*
Iowa casino financial data, 41*t*
Las Vegas downtown gaming revenue, by type, 29*t*
Las Vegas strip gaming revenue, by type, 28*t*
lottery contributions to beneficiaries, by state, 83*t*–84*t*
lottery income/apportionment of funds, by state, 74*t*
Louisiana land-based casino financial data, 37(*t*4.14)
Louisiana riverboat casino employment data, 62(*t*6.3)
Louisiana riverboat gaming statistics, 36*t*
Louisiana video gaming statistics, 38(*t*4.15)
Maryland Problem Gambling Helpline, callers to by gender, 69(*f*6.5)
Maryland Problem Gambling Helpline, callers to by identity of caller, 70(*f*6.7)
Maryland Problem Gambling Helpline, callers to by race/ethnicity, 70(*f*6.6)
Maryland Problem Gambling Helpline, phone/text/online chat contacts received by, 68*f*
Maryland Problem Gambling Helpline, problems reported by callers to, 69(*f*6.4)
Mississippi casino revenues, by region, 34*t*
Mississippi casino tables hold, by game/region, 35*t*
Missouri casino gaming financial data, by casino location, 39(*t*4.19)
Missouri casinos, arrests at by month, 71(*t*6.5)
Nevada casino industry employment data, 62(*t*6.2)
Nevada gaming revenue, 26*f*
Nevada gaming revenue, by type, 27*t*
Nevada sports betting, amount wagered on by sport, 95*t*
Nevada sports betting financial data, 88*t*
Nevada sports betting revenue percentages, by sport, 97*t*
Nevada sports betting revenues, by sport, 96*t*
New Jersey casino gaming revenue, 30(*t*4.4)
New Jersey casino gaming revenue, by type, 31*t*
New Jersey online gambling revenues, 104*f*
North American Thoroughbred racing, total purse amounts, 92*f*
North American Thoroughbred racing, total wager amounts, 91*t*
Pennsylvania casino revenues, 32(*t*4.8)
public opinion on selected issues, 7*f*
riverboat casino financial data, by state, 37(*t*4.13)
South Dakota casino revenues, 44*f*
tax revenues generated by casino gambling, by state, 15*t*
Texas lottery game, percentage of people who played, 80*f*
Texas lottery players *vs.* nonplayers, demographic characteristics of, 79*t*–80*t*

Texas state lottery tickets, median spending on, 82*t*
tribal gaming revenues, by region, 50*t*
tribal gaming revenues, growth in, 46*f*
tribal gaming revenues, growth in, by region, 50*f*
tribal gaming revenues, nationwide, 49*f*
West Virginia casino revenues, 42(*f*4.2)
winnings reported to IRS, 14(*t*2.3)
Statute of Anne, 2
Stewart, David O., 101–102, 106
Stodghill, Ron, 75
Strauss, Valerie, 84
Strip. *See* Las Vegas, Nevada
Suicide, casino gambling and, 66
"Suicide Ideations, Suicide Attempts, and Completed Suicide in Persons with Pathological Gambling and Their First-Degree Relatives" (Black et al.), 66
Summa Corporation, 4

T

Taggart, William, 51
Tagliabue, Paul, 87
Takeout
of horse racetrack operators, decline of, 91
pari-mutuel gambling, 89
"Taking the Mystery out of the Machine: A Guide to Understanding Slot Machines" (AGA), 22
Tampering, sports, 99–100
Taxes
federal excise on sports wagers, 94
gambling, 14
government revenues from gambling, 59
lottery, economic results of, 81
from lottery winnings, 84
paid by casinos, 58
regressive lottery, 78
revenues generated by online gambling, 107
on state lotteries, 76
tax revenues generated by casino gaming, by state, 15*t*
Taylor, Eric, 20
Taylor, Paul
on gamblers, 15–16
on problem gamblers, 17
on public opinion on gambling, 57
Tetreault, Steve, 102
Texas
demographics of lottery players in, 78–79
lottery, Dallas Cowboys–themed scratch-off of, 76
lottery game, percentage of people who played, 80*f*
lottery players *vs.* nonplayers, demographic characteristics, 79*t*–80*t*
median spending on lottery tickets, by demographic characteristics of players, 82*t*
unclaimed lottery winnings, use of, 83–84
Texas Lottery, 84–85
"Texas Lottery: A Different Game Than State Was Sold Two Decades Ago" (Dexheimer), 84–85
Texas Lottery Commission, 78–79
Theatrical Market Statistics, 2015 (MPAA), 58
Theatrical Market Statistics, 2016 (MPAA), 9
Thies, Clifford F., 66
Thomas, Trang, 19
Thoroughbred horse racing. *See* Horse racing
Thoroughly Thoroughbred (Jockey Club), 89
Thunder Valley Casino, 51
Thygesen, Kylie, 66
"To Slot Players, Palms' Sideshow a Freak Show" (Benston), 24
Toensing, Gale Courey, 60
Tohono O'odham Nation, 52
Toneatto, Tony, 19
Tote board, 90
Tourism
Atlantic City, 64–65
casinos and, 63–64
Las Vegas, 64
"Track Information" (U.S. Trotting Association), 90
Trans America Wire, 98
Treatment, gambling disorder
overview of, 67–68
treatment methods, 19–20
treatment organizations, 19
Tribal gaming
tribal casinos, revenues generated by, 9
tribal governments, gambling revenues generated for, 15
U.S. Supreme Court on, 5
See also Casinos, Native American tribal
Tribal-state compacts
of California, 55
description of, 46
state regulation of tribal casinos via, 47
"Tribes Clash as Casinos Move Away from Home" (Lovett), 52
Triple Crown, 90
Trump, Donald, 52
"Trump Administration Faces Test of Off-Reservation Gaming Policy" (Indianz.com), 52
Tubbs, Alice Ivers, 3, 43
Tulane University, 63
Tunica County, Mississippi, 61
20th/21st centuries, 3
21st Annual Report to the Louisiana State Legislature (Louisiana Gaming Control Board), 36
2013 State of the States: The AGA Survey of Casino Entertainment (AGA)
on acceptability of gambling, 57
on casino acceptability, 22
on casino games, popularity of, 22
on casino visitors, activities participated in by, 23–24
on casino visitors, numbers of, 63
on gamblers, 16
on gamblers budget limits, 67
on Internet wagering, numbers participating in, 106–107
on lottery tickets, number of purchasers, 78
2016 Annual Report (Iowa Racing and Gaming Commission), 41
2016 Annual Report (Kansas Racing and Gaming Commission), 43
2016 Annual Report (MGM Resorts International), 10
2016 California Tribal Gaming Impact Study: An Economic, Fiscal, and Social Impact Analysis with Community Attitudes Survey Assessment (California Nations Indian Gaming Association)
on economic output of California tribal casinos, 59
on tribal gaming, economic impact on state, 56
"2016 Indian Gaming Revenues Increase 4.4 Percent" (NIGC), 45, 50
"2016 Las Vegas Year-to-Date Executive Summary" (LVCVA), 64
"2016 Monthly Casino Revenue Report" (Ohio Casino Control Commission), 42
2016 State of the States: The AGA Survey of the Casino Industry (AGA)
on California tribal casinos, 55
on commercial casino revenues, 21, 45, 58
on commercial casino tax revenues, 59
on commercial casinos, numbers operating in U.S., 25
on commercial casinos, revenues generated by, 9
on commercial casinos operating in Nevada, 25–26
on Illinois casino market, 36
on Louisiana casino market, 36
on Mississippi Gulf Coast casino market, 34
on numbers employed in commercial casinos, 59–60
on states allowing commercial casinos, 4

U

UIGEA. *See* Unlawful Internet Gambling Enforcement Act
"Unclaimed Prizes" (California State Lottery), 84
Underage gambling
in casinos, 68
Internet, 109–110

Unions, tribal casinos and, 63
United Auburn Indian Community, 51
United Farm Workers, 55
United States
Americans who gambled in previous 12 months, by gambling type, 7*t*
gambling in, overview of, 1
Internet gambling in, 5–6, 101
issues concerning/social impact of gambling, 6
Native American tribes, federal recognition of, 45, 47–48
1900, gambling in since, 3–6
19th century, gambling in during, 2
online gambling, growth of in, 104–105
precolonial/colonial era, gambling in during, 1–2
public opinion on gambling in, 6–7, 7*f*
of U.S. lotteries, future of, 86
United States Online Gaming: Monthly Statewide and National (Center for Gaming Research), 104
Unlawful Internet Gambling Enforcement Act (UIGEA)
description of, 5
effects of, 102
fantasy sports exemption of, 96
passing of, 11
U.S. Census Bureau
on Native Americans/Alaskan Natives, 45
on state lottery revenues, 73
U.S. Commission on the Review of the National Policy toward Gambling, 4
U.S. Department of Justice (DOJ)
on Internet gambling, 5–6
Internet gambling crackdown/reversal of, 102
on Internet gambling, states regulation of, 12
U.S. Department of Labor, 60
U.S. film industry, 58
U.S. Government Accountability Office (GAO)
on legality of tribal casinos, 46
on tribal recognition process, 48
U.S. Senate, 94
U.S. Small Business Administration, 50–51
U.S. Supreme Court
California v. Cabazon Band of Mission Indians, 46
Michigan v. Bay Mills Indian Community, 52
sports gambling, agreement to hear arguments on, 94
on tribal gaming, 5
"U.S. Supreme Court to Review N.J. Sports Betting Case; Supporters React" (Brennan), 94
U.S. Trotting Association, 90
U.S. Census Bureau, 21

V

"Validating the Gambling Functional Assessment—Revised in a United Kingdom Sample" (Weatherly et al.), 17
Varghese, Romy, 32
Vaughn, Chris "BluffMagCV," 108
"Vegas Goes for Edgier Ads" (Howard), 64
"Vegas Tourism Companies Embrace Social Media Strategies" (Finnegan), 64
Velotta, Richard N., 64
Vick, Michael, 99
Victoria, Queen of England, 3
Video lottery terminals
description of, 75
Maryland, 42–43
Video slot machines, 22
Vigorish, 94
Violations, tribal casino, 47
Voluntary Exclusion Program, Missouri, 67

W

Wagering. *See* Betting
Wages
employment in tribal casinos, effects on local economies, 61, 63
median annual, gaming service occupations *vs.* all occupations, 60*f*
paid by commercial casinos, 60
Walker, Douglas M.
Casinonomics: The Socioeconomic Impacts of the Casino Industry, 59, 61
"Casinos and Political Corruption in the United States: A Granger Causality Analysis," 69
"Do Casinos Really Cause Crime?," 65
Washington, George
lottery of, 73
Maryland Jockey Club, membership in, 89
Weatherly, Jeffrey N., 17
Weaver, Jay, 47
Wells, Barbara, 51
Welty, Ward R., 66
West Virginia
casino revenues of, 42(*f*4.2)
commercial casinos of, 40
Westphal, James R., 3
"What Happens Here, Stays Here" (LVCVA ad campaign), 64
"What We Do" (BIA), 45
Whether Proposals by Illinois and New York to Use the Internet and Out-of-State Transaction Processors to Sell Lottery Tickets to In-State Adults Violate the Wire Act (DOJ)
on Internet gambling, 5–6
on Internet gambling, states regulation of, 12
online lottery sales and, 102
Wild Card (lotto game), 77
Winnings, unclaimed lottery, 82–84
Wire Act
lottery ticket sales, online and, 5, 12
online lottery sales, allowance for, 102
states, legal Internet gambling in, 107
Withholdings, lottery winnings, 84
Without Reservation: How a Controversial Indian Tribe Rose to Power and Built the World's Largest Casino (Benedict), 53
Women, problem gambling among, 68
Wong, Gloria, 16
Wong, Paul W. C., 66
World Series of Poker (WSOP), 101
World War II
Las Vegas, postwar casino growth in, 4
Mississippi, gambling in during/after, 33
Worldwide online gambling, 103–140
WSOP (World Series of Poker), 101
Wynn, Steve, 58
Wyoming Lottery, 86

Y

"Year End Summaries for 2004–2010" (LVCVA), 64
Yearlings, horse, 99
Young people
gambling among, 16–17
online gambling among, 107
sports wagering among, 97–98, 99–100
underage gambling, 109–110
See also Adolescents

Z

Zinke, Citizens against Reservation Shopping v., 52
Zinke, Ryan, 52

CPSIA information can be obtained
at www.ICGtesting.com
Printed in the USA
FFOW01n2113130518
46624902-48683FF